Handbook on SPLINES for the User

Handbook on SPLINES for the User

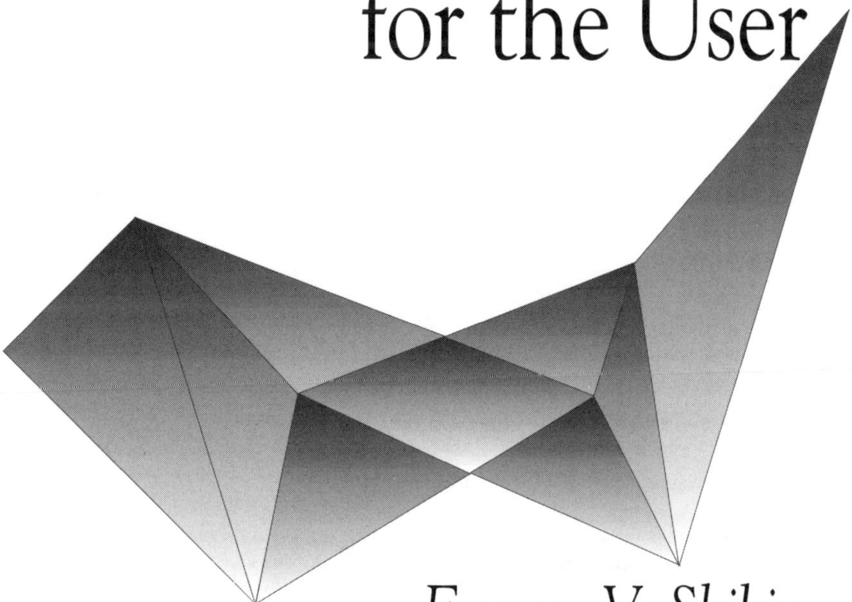

Eugene V. Shikin

Alexander I. Plis

CRC Press

Boca Raton New York London Tokyo

Library of Congress Cataloging-in-Publication Data

Catalog record is available from the Library of Congress.

Direct all inquiries to CRC Press, Inc., 2000 Corporate Blvd., N.W., Boca Raton, Florida 33431.

International Standard Book Number 0-8493-9404-X
Printed in the United States of America 1 2 3 4 5 6 7 8 9 0
Printed on acid-free paper

To our parents

Nadezhda I. Shikina

Fedora D. Plis and Ivan B. Plis

Preface

The paths connecting the initial intentions of the authors and the published book every so often turn out to be so unique that their description deserves a special account which is hardly interesting for readers. Besides, the pleasure which the story-teller usually finds in the account can not overshadow the main event — the actual publishing of the book.

The coming out of a book provides its authors with a worthy cause for happiness. It should be remarked, however, that the happiness becomes complete only after the last copy of the published book finds its reader for whom the book is both necessary and useful. Completing the work on the manuscript the authors hope to experience this happiness as well.

We sincerely hope that this happiness will be shared along with us by all those contributing to the publishing of the book. In addition to the publishing house and the authors placed on its cover, there exist others participating in the event. None of the authors are able to overestimate their combined efforts in the publishing of the handbook.

We consider it to be our pleasant duty to give their names and express our heartfelt gratitude to Sergey V. Ezhkov, Anna L. Kirin, Andrey F. Mubarakshin, Alexander D. Pankratov, Mikhail Yu. Spokoinyi, and Mikhail M. Frank-Kamenetskii.

Moscow
April 1995

Eugene V. Shikin
Alexander I. Plis

Contents

Why splines?

At the foundation of every creation is the idea of assembling the whole from the carefully selected parts. As old as our civilization, this idea is still a long way from the ultimate resolution. This is quite natural: to break something is much simpler than to scrupulously collect and glue together the fragments, and to do it in such a way that the creation is a pleasure for the eye.

Throughout its history, mankind tried (and successfully enough at times) to solve the problem. Here we shall give only two examples choosing the ones closest to the problems, which the present handbook aims to solve.

There exists a beautiful legend describing the origin of the *Bukhara china*. In the ancient times, chinaware was produced only in China. The price of china was very high, and because of this it was especially valued by the sovereigns of the lands large and small.

A splendid china service was made to order of the Bukhara emir and transported to his court. But despite all the precautions, the service did not withstand the caravan journey of many days and a major part of the really fragile chinaware was broken. In the present twentieth century we can not fully measure the subjects' experience of fear of the emir. But they knew sufficiently well the consequences of the emir's anger. It was necessary to quickly find an adequate way out of the situation. And, as the legend says, the way was found. Goldsmiths, famous throughout the Middle East, fastened the shivers of bowls, dishes, saucers, soup-, dinner-, side-plates, and cups together using the thinnest plates. The skillful jewelers not only recreated the elegant beauty of the chinaware, but returned functionality to them as well — the recreated products could be used for their original purpose. The performed operations imparted a uniqueness to the long-suffering service so attractive that later the chinaware ordered from China was accepted only after a peculiar dismemberment and the subsequent "gold" assembly. Today, travelling through Middle Asia, one can still encounter china or faience teapots repaired in a similar ("Bukhara") fashion, but without the gold plates.

And here is one more well known *shipbuilding* example.

A curious mankind, attracted by the surrounding expanse of water did not use the gouged boats and rafts too long, for their nautical qualities were unsatisfactory. The yearning for unknown lands beyond the sea was of considerable importance in searching for more and more perfect means of conveyance over the sea. As was often the case of a great discovery, an idea to create special sea-going ships arose and began to be independently developed in diverse parts of the world (true, with various success). The key to the idea was a realization of the necessity to begin building the ship with rigidly built framework. With time this fundamental problem was solved by the universal trial-and-error method, and quite satisfactory shapes giving buoyancy, stability, and maneuverability to the ships were found. And shortly after that mankind swam in the epoch of great discoveries aboard the caravelles, clippers, and schooners.

Later these ideas were used successfully not only in shipbuilding but in motorcar construction and aircraft building as well as in architecture.

Note that despite the distinction of the materials used in both cases, the product was recreated according to a given template by fitting the available fragments. True, in solving the china problem the creative group was divided into two independent parts, one of which (Chinese) created and the other (Bukharian) recreated the practically ready standards (there existed also the third part to break). The shipbuilders would be forced to make both the wooden skeleton structure and the shell themselves (it must be emphasized that woods occupied half the Earth at that time).

Let us try to translate the considered problem into the geometric language.

Geometric model. After a description of the desired curve or surface by the array of the support points chosen correspondingly, construct the curve or surface by the points through the choice of necessary fragments.

Solving the problem of the construction of the curve or surface is divided into the following steps:

1) determination of a suitable set of support points,

2) development of a universal and relatively simple method for constructing the elementary fragments of the curve or surface,

3) search for an effective mechanism for a smooth joining of the elementary fragments, and

4) analysis of the results.

It is clear that the solution of the problem is not unique, since the given array of the points may be approximated by the smoothing curve or surface in a number of ways. Therefore, a set of the approximate geometric objects is practically boundless. At the same time, when solving specific problems it is necessary, as a rule, to find only one approximating, sufficiently well-behaved object. For this reason it would be prudent to first understand clearly what object plays the role of a needle in a haystack of possibilities.

The selection of the elementary fragments (segments of curves or patches of surfaces) needs to be made so that their description will be simple, universal, and geometrically clear. The advantage of simplicity is sufficiently obvious. The universality permits the volume of calculations to be relatively small. The clarity of the geometric characteristics is inestimable in joining the fragments and the subsequent analysis of the resulting curve or surface. In many cases the geometric continuity, that is, the smooth variation of the curve's tangent and curvature or the surface's tangent plane and curvature, respectively, is usually sufficient. If one wants to use the scalar polynomial functions for the fragment description, then the above continuity can be obtained by the polynomials of the third order (in case of surface, with respect to either of the two auxiliary variables).

The problem of the uniqueness of the approximation problem plays a role of no small importance.

Of course, meeting all of the above restrictions at the same time is not simple. Using the well known classes of curves or surfaces (for example, given by polynomials or piecewise linear functions), one can meet only a part of the requirements, though, in each case its own. Because of this, the desire to combine most advantages of these classes in some successful way is reasonable.

Let us describe some of the requirements which need to be imposed on the selected classes of curves and surfaces and the fulfillment of which is necessary for successfully resolving the approximation problem (needless to say others exist):

1. The selected class must be described relatively simply.
2. The curves or surfaces included in the selected class should not have any singularities, that is, they must be sufficiently smooth (for example, not have a break, or have a tangent changing continuously, or a continuous curvature).
3. Search for the desired curve or surface in the selected class must be relatively simple, assuming the availability of the effective algorithm of its construction.
4. For the sufficiently large arrays of the points, the constructed curves or surfaces must behave quite predictably.

It is interesting to note that curves and surfaces satisfying the above requirements exist. Each of them consists of smooth fragments adjusted in concord. It is agreed that these curves and surfaces are called *splines*.

By this means the inevitability of the spline advent flew through the air over a long period of time. Great Gogol was able to express this general feeling in a most intelligible and at the same time poetical form in his famous play "Marriage". A young lady while choosing her would-be husband says:

"If we were to take Nikolai Ivanovich's lips and place them under the nose of Ivan Kusmich, take a little of Baltazar Baltazarovich's looseness, and probably then add the stoutness of Ivan Pavlovich — *I would be resolved at once.*"

The world of splines is extremely various and complex. It is not easy to enter this world. We propose to use for this purpose one of the most natural and simple ways — *three-order-splines*, that is, splines given by polynomials of the third degree (cubic and bicubic).

On the structure of the handbook

During the preparation of this handbook, the authors proceeded on the assumption that the material enclosed in it may be useful to those making their acquaintance with splines, as well as to those who have already used splines in their work. At the same time, this book is not intended to be a textbook, and it would be advisable to refer to the appropriate books and scientific papers for comprehensive explanations and detailed proofs (a small number is given at the end of the book).

The aim of the authors was to create a handbook that was conceptually simple and accessible to a wide circle of users providing each of them with an opportunity to learn to effectively solve the basic problems "on splines" and with an introduction to the methods used.

At all stages of our work on the handbook, a lot of attention was given to the development of its structure and the presentation of the material most convenient to the user. It would be worthwhile to describe in more detail some of the distinctive features of this handbook.

One of the basic ideas which the authors have attempted to imbue into the book is helping the user to choose precisely the method which is best suited for solving his specific problem. For this realization the authors use a comparison of various approaches and methods as well as a description of their peculiarities as the constant component of the book's outline. This aim is also served by adherence to the historical and/or the logic chronology of the emergence of the methods contained in the book, and the successive description of how the ideas and advantages of the approaches found earlier multiply with their modification and evolution and the developing methods gradually rid themselves of the disadvantages.

Opening the handbook at an arbitrary page, a user has the right to expect a relatively easy understanding of the material described without studying the whole book. The authors have tried to meet this expectation during development of the handbook's structure and the selection of the material. But in a large number of cases, the sequence of the material to be reviewed is sufficiently rigidly determined. For example, if a reader wants to make an acquaintance with the interpolating spline functions of two variables, it would be appropriate first to review the section where the interpolating spline functions of one variable are described. In perfect analogy, it would be prudent to forestall the study of the subsection of Beta-spline surfaces with a review of the corresponding one-dimensional subsection where the Beta-spline curves are concerned.

We would like to emphasize that the chapters devoted to the spline functions of two variables (Chapter 2) and the spline surfaces (Chapter 4) are far more complex than the chapters where the spline functions of one variable (Chapter 1) and the spline curves (Chapter 3) are considered. This is natural, for the objects are more complex by themselves. For this reason, to somewhat simplify the use of the handbook, the authors tried to choose the spline classes so that practically every one-dimensional spline class from the handbook would have a two-dimensional analogy in it, and, in turn, every two-dimensional spline would lean upon a one-dimensional fore-runner.

The presentation of the material on the whole is constructed in such a way that every chapter, and, by and large, every subsection may be read independently of the others. In addition, in order to help the user to better orient himself in the pages of the book, the authors used similar descriptions of the construction of various splines as well as a minimum of cross-references.

In fact, the handbook offered is divided into four subhandbooks with relatively independent and explicitly chosen subjects. A chapter is devoted to every subhandbook. It is the authors' opinion that this will help the user receive the answers on the questions of interest to him without gathering information throughout the whole book. Wherever the materials of the other chapters are appropriate (or even necessary) for use, precise references are given.

The material was selected in order to enable a reader to get an introduction to the basic notions, facts, and methods and to enter relatively easily into world of splines.

Many of the results obtained by the calculations possess certain peculiarities: sometimes it is sufficiently difficult to explain why at this point of calculations one must use this approach and not another, or why a certain sequence of steps is to be used in a nonstandard situation, etc. Here, intuition and experience play an important role. As for intuition, the situation is not always simple. Experience however may be gained in a number of ways. The present handbook is just one of these ways.

Constrained by the realization of the responsibility to the readers rather than by the volume, a decision was made not to focus their attention on a variety of the details in the handling of splines, since we understand all too well how entangled in them one may become. For this reason we carefully chose and rigidly structured the selected material so that the user's view of the spline tool and its possibilities be right in general and the user would be able to solve the basic problems of approximation sufficiently fluently.

In addition, we tried to construct the contents in such a way so that in using the material of the book a reader would have a chance to test his understanding.

The book is divided into two large parts, one of which is devoted to spline functions and the other to geometric splines. More simply, one may say that calculations predominate in the first part, whereas the leading role belongs to pictures in the second part. But the two parts are closely interrelated through the commonness of the basic ideas. One of the connecting ideas is the computer implementation of the spline algorithms.

An essential part of the handbook is SplineGuide (on applied diskette) containing programs to build interpolating and smoothing splines of all the classes described in Part I (spline functions of one and two variables) and in Part II (spline curves and spline surfaces).

Part I
Spline functions

In practice many of the constructed curves or surfaces have a sufficiently complex shape which does not permit a universal analytic description at large with the help of the elementary functions. For this reason the objects are assembled out of the relatively simple smooth fragments — segments (for curves) and patches (for surfaces), each being represented as a graph of the elementary function of one or two variables.

For the resulting composite curve or surface to be sufficiently smooth, one must be particularly attentive to the joining sites.

It is natural to require the smooth functions graphs used in the construction of the partial curves and surfaces to have a similar nature, for example, be polynomials of the same degree. The last is selected from the simple geometric reasoning and, as a rule, is not at all high: for a tangent that changes smoothly along the whole composite curve it is sufficient to describe the joining curves by the cubic polynomials. Coefficients of the polynomials may always be sorted out in such a way that the corresponding composite function has the continuous second derivative.

It is quite reasonable to use the results of solving the one-dimensional problems in the construction of the corresponding composite surfaces. This is how the bicubic spline functions (the polynomials of the third degree with respect to either of the two variables) appear. Of course, the handling of these splines requires a vastly larger volume of calculations.

Thus in Part I, the smooth curves and surfaces represented as the graphs of the piecewise polynomial functions of a lower degree are considered. It is precisely this class of curves, the construction of which has begun relatively recently, that provides the impetuous history of the one-dimensional (and further the two-dimensional) splines.

Chapter 1

Spline functions of one variable

Let on segment $[a, b]$ be given a *grid* ω

$$a = x_0 < x_1 < \ldots < x_{m-1} < x_m = b.$$

Points x_0 and x_m are called *end (boundary) knots* of grid ω, and points $x_1, \ldots, x_{m-1}$ — its *inner knots* (Figure 1). The grid is called *uniform* if a distance between every two neighboring knots is the same.

Function $S(x)$ given on segment $[a, b]$ is called a *spline of type* $p+1$ *(degree* p*)* if this function

1) is a polynomial of given degree $p \geq 2$ on every segment

$$\Delta_i = [x_i, x_{i+1}], \quad i = 0, 1, \ldots, m-1,$$

that is, it may be written in the form

$$S(x) = S_i(x) = \sum_{k=0}^{p} a_k^{(i)}(x - x_i)^k, \quad i = 0, 1, \ldots, m-1,$$

and

2) is $p-1$ times continuously differentiable on segment $[a, b]$; that is, $S(x) \in C^{p-1}[a, b]$.

Remark. Index (i) of coefficient $a_k^{(i)}$ indicates that the system of numbers defining function $S(x)$ is of its own for every partial segment Δ_i.

Spline $S(x)$ is a polynomial of degree p on every segment Δ_i, $i = 0, 1, \ldots, m-1$, and is defined on the segment by $p+1$ coefficients. A total number of the partial segments is equal to m; that is, to completely define the spline it is necessary to find $(p+1)m$ numbers

$$a_k^{(i)}, \quad k = 0, 1, \ldots, p, \quad i = 0, 1, \ldots, m-1.$$

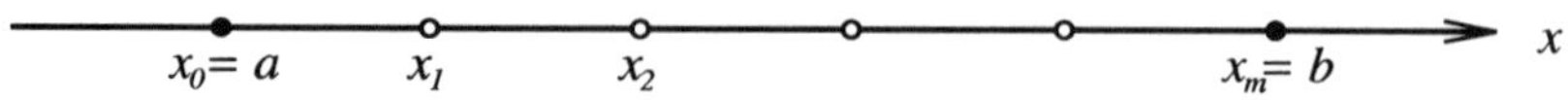

Figure 1.

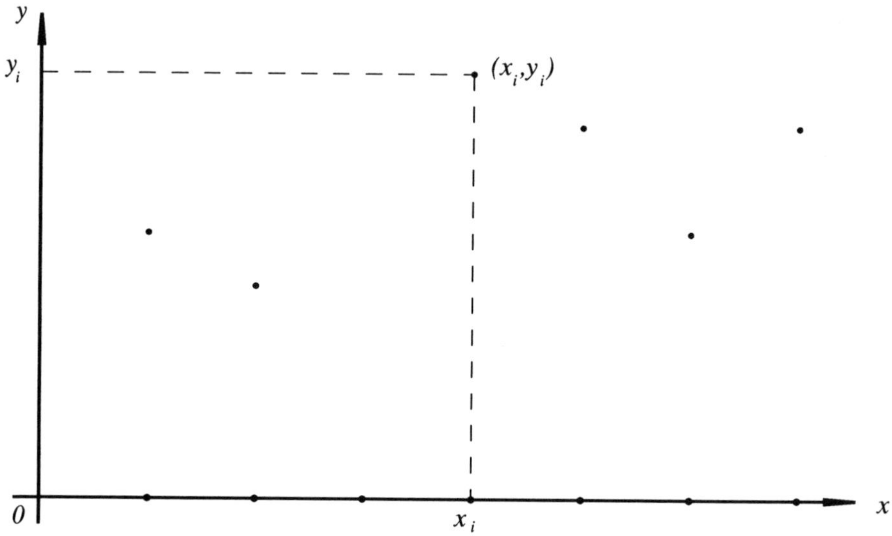

Figure 2.

Condition $S(x) \in C^{p-1}[a,b]$ means the continuity of function $S(x)$ and its derivatives $S'(x)$, $S''(x)$, ..., $S^{p-1}(x)$ at all inner knots of grid ω. A total number of such knots is equal to $m-1$; that is, we have $p(m-1)$ conditions (equations) for the coefficients of all the polynomials.

For a complete definition of the spline we need $(p+1)m - p(m-1) = m+p$ conditions (equations). A choice of the additional conditions is defined by the character of the considered problem or simply by a desire of the user.

The problems of interpolation and the smoothing most often considered are those that are required to construct one or the other of the splines by a given array of the points on plane (x_i, y_i), $i = 0, 1, \ldots, m$ (Figure 2).

In the *interpolation problem* it is necessary that the spline's graph passes through the points (x_i, y_i), $i = 0, 1, \ldots, m$. The requirement imposes $m+1$ additional conditions (equations) on its coefficients. The remaining $p-1$ conditions (equations) for a unique definition of the spline are most commonly given as the values of the lowest derivatives at the ends of the segment $[a,b]$ — *end (boundary) conditions.* The possibility of the choice of various boundary conditions causes a wide variety of the interpolating spline types and their properties.

In the *smoothing problem* the spline is constructed in such a way that its graph goes near points (x_i, y_i), $i = 0, 1, \ldots, m$, rather than through them. A measure of the nearness may be defined differently, resulting in a wide variety of smoothing spline types.

The described possibilities of choosing the spline function construction do not exhaust their variety. Initially the piecewise polynomial spline functions were considered. As the area of their application began to widen, the splines with useful and interesting properties composed of the other elementary functions came into existence.

1.1 Interpolating cubic splines

1.1.1 Statement of the interpolation problem

Let a grid ω

$$a = x_0 < x_1 < \ldots < x_{m-1} < x_m = b$$

be given on the segment $[a, b]$. Consider a set of numbers

$$y_0, y_1, \ldots, y_{m-1}, y_m.$$

Problem. Find a smooth (on the segment $[a, b]$) function $\sigma(x)$, the values of which coincide with given numbers at the knots of the grid ω, that is

$$\sigma(x_i) = y_i, \quad i = 0, 1, \ldots, m-1, m.$$

Remark. The formulated interpolation problem consists in the reconstruction of a smooth function given by a table. It is clear that the problem has a set of various solutions. With certain additional conditions for the desired function, the function may be constructed uniquely.

The problem of approximation of a function defined by the formula

$$y = f(x), \quad x \in [a, b],$$

with the help of a function with sufficiently good properties is widely met in applications. For example, such a problem arises if the calculation of the given function's values $f(x)$ at the points of the segment $[a, b]$ involves substantial difficulties and/or the given function $f(x)$ is not of required smoothness, and we need to use another function which closely approximates the above function and has none of its weaknesses.

Problem of the function interpolation. Construct a smooth (on the segment $[a, b]$) function $\sigma(x)$ which coincides with the given function $f(x)$ at the knots of the grid ω

$$\sigma(x_i) = f(x_i), \quad i = 0, 1, \ldots, m-1, m.$$

1.1.2 Definition of the interpolating cubic spline function

A function $S(x)$ defined on the grid ω is called *an interpolating cubic spline function* if the function

1) is a cubic polynomial

$$S(x) = S_i(x) = a_0^{(i)} + a_1^{(i)}(x - x_i) + a_2^{(i)}(x - x_i)^2 + a_3^{(i)}(x - x_i)^3$$

on each partial segment $[x_i, x_{i+1}]$, $\quad i = 0, 1, \ldots, m - 1$,

2) has the second continuous derivative on the segment $[a, b]$, that is, the function is of class $C^2[a, b]$, and

3) satisfies the conditions

$$S(x_i) = y_i, \quad i = 0, 1, \ldots, m. \tag{$*$}$$

The spline function $S(x)$ is a cubic polynomial on each partial segment $[x_i, x_{i+1}]$, $i = 0, 1, \ldots, m - 1$, and hence it is defined on the segment by four coefficients. A total number of the partial segments is equal to m. Consequently, when we find $4m$ numbers

$$a_0^{(i)}, a_1^{(i)}, a_2^{(i)}, a_3^{(i)}, \quad i = 0, 1, \ldots, m - 1$$

the desired spline function will be defined.

A condition $S(x) \in C^2[a, b]$ means a continuity of the function $S(x)$ and its derivatives $S'(x)$ and $S''(x)$ at all inner knots of the grid ω. The number of the inner knots is equal to $m - 1$. Therefore, we have $3(m - 1)$ conditions (equations) for the desired coefficients. In view of the conditions $(*)$ the total number of the conditions (equations) is equal to $3(m - 1) + (m + 1) = 4m - 2$.

1.1.3 End (boundary) conditions

Two more conditions are given in the form of restrictions on the spline values and/or the values of its derivatives at the end points of the segment $[a, b]$. Four of the most commonly used boundary conditions are described below.

End conditions of the first type

The values of the first derivative are given at the end points of the segment $[a, b]$

$$S'(a) = f'(a), \quad S'(b) = f'(b).$$

End conditions of the second type

The values of the second derivative are given at the end points of the segment $[a, b]$

$$S''(a) = f''(a), \quad S''(b) = f''(b).$$

End conditions of the third type

The respective values of the first and second derivatives are equal at the end points of the segment $[a, b]$

$$S'(a) = S'(b), \quad S''(a) = S''(b).$$

The conditions are called *periodic.*

Comment. The fulfillment of the conditions is naturally required when the interpolated function is periodic and its period is equal to $T = b - a$.

End conditions of the fourth type

The conditions

$$S'''(y, x_1 - 0) = S'''(y, x_1 + 0), \quad S'''(y, x_{m-1} - 0) = S'''(y, x_{m-1} + 0)$$

require a special comment.

Comment. In general, the third derivative of the spline function $S(x)$ is discontinuous at the inner knots of the grid. But the number of its breaks may be lowered using end conditions of the fourth type. In this case the constructed spline function will be continuously differentiable three times on the segments $[x_0, x_2]$ and $[x_{m-1}, x_m]$.

1.1.4 Construction of the interpolating cubic spline function

We describe a calculation method where the number of desired values is equal to $m + 1$ instead of $4m$.

In each segment $[x_i, x_{i+1}]$, $\quad i = 0, 1, \ldots, m - 1$, the interpolating spline function is sought in the following form

$$S(x) = S_i(x) =$$

$$y_i(1-t)^2(1+2t) + y_{i+1}t^2(3-2t) + n_i h_i t(1-t)^2 - n_{i+1}t^2(1-t), \quad (**)$$

where

$$h_i = x_{i+1} - x_i, \quad t = \frac{x - x_i}{h_i}$$

and numbers n_i $(i = 0, 1, \ldots, m)$ are a solution of the linear algebraic system the form of which depends on the end condition type.

For the *end conditions of the first and second types*, the system takes the form

$$\begin{cases} 2n_0 + \mu_0^* n_1 = c_0^*, \\ \lambda_i n_{i-1} + 2m_i + \mu_i n_{i+1} = c_i, \quad i = 1, 2, \ldots, m-1, \\ \lambda_m^* n_{m-1} + 2n_m = c_m^*, \end{cases}$$

where

$$c_i = 3\left(\mu_i \frac{y_{i+1} - y_i}{h_i} + \lambda_i \frac{y_i - y_{i-1}}{h_{i-1}}\right).$$

Coefficients $\mu_0^*, c_0^*, \lambda_m^*, c_m^*$ depend on the choice of the end conditions.
The end conditions of the first type:

$$\mu_0^* = 0, \quad c_0^* = 2y_0',$$

$$\lambda_m^* = 0, \quad c_m^* = 2y_m'.$$

The end conditions of the second type:

$$\mu_0^* = 1, \quad c_0^* = 3\frac{y_1 - y_0}{h_0} - \frac{h_0}{2}y_0'',$$

$$\lambda_m^* = 1, \quad c_m^* = 3\frac{y_m - y_{m-1}}{h_{m-1}} + \frac{h_{m-1}}{2}y_m''.$$

For *the end conditions of the third type*, the linear system for the desired numbers n_i, $i = 1, 2, \ldots, m$, may be written as follows:

$$\begin{cases} 2n_1 + \mu_1 n_2 + \lambda_1 n_m = c_1, \\ \lambda_i n_{i-1} + 2n_i + \mu_i n_{i+1} = c_i, \quad i = 2, \ldots, m-1, \\ \mu_m n_1 + \lambda_m n_{m-1} + 2n_m = c_m. \end{cases}$$

As a consequence of the periodicity condition, $n_0 = n_m$, the number of the unknowns in the last system is equal to m.

For *the end conditions of the fourth type*, the linear system for the desired numbers n_i, $i = 1, 2, \ldots, m$, may be written as follows:

$$\begin{cases} (1 + \gamma_0)n_1 + \gamma_0 n_2 = c_1^*, \\ \lambda_i n_{i-1} + 2n_i + \mu_i n_{i+1} = c_i, \quad i = 2, \ldots, m-2, \\ \gamma_m n_{m-2} + (1 + \gamma_m)n_{m-1} = c_{m-1}^*, \end{cases}$$

where

$$\gamma_0 = \frac{h_0}{h_1}, \quad c_1^* = \frac{1}{3}c_1 + 2\gamma_0 \frac{y_2 - y_1}{h_1},$$

$$\gamma_m = \frac{h_{m-1}}{h_{m-2}}, \quad c_{m-1}^* = \frac{1}{3}c_{m-1} + 2\gamma_m \frac{y_{m-1} - y_{m-2}}{h_{m-2}}.$$

If the system solution is found, the numbers n_0 and n_m may be calculated by the formulas

$$n_0 = 2\left(\frac{y_1 - y_0}{h_0} - \gamma_0^2 \frac{y_2 - y_1}{h_1}\right) - (1 - \gamma_0^2)n_1 + \gamma_0^2 n_2,$$

$$n_m = 2\left(\frac{y_m - y_{m-1}}{h_{m-1}} - \gamma_m^2 \frac{y_{m-1} - y_{m-2}}{h_{m-2}}\right) - (1 - \gamma_m^2)n_{m-1} + \gamma_m^2 n_{m-2}.$$

Important remark. The matrices of all three linear algebraic systems are matrices with the diagonal predominance. Such matrices are nondegenerate, and, as a result, each of the systems has a unique solution.

Theorem. An interpolating cubic spline function satisfying the relations $(*)$ and the end condition of one of the four types above exists and is unique.

Therefore, to construct the interpolating cubic spline function it is necessary to find the numbers $n_0, n_1, \ldots, n_m$.

When the numbers are found, a spline value $S(x)$ at an arbitrary point of the segment $[a, b]$ may be calculated by formula $(**)$. But in practice, the following algorithm for the determination of the values $S(x)$ is more convenient.

Let $x \in [x_i, x_{i+1}]$. First the numbers A and B are calculated using formulas

$$A = -2\frac{y_{i+1} - y_i}{x_{i+1} - x_i} + n_i + n_{i+1}, \quad B = -A + \frac{y_{i+1} - y_i}{x_{i+1} - x_i} - n_i,$$

then we calculate the value of the spline-function $S(x)$ at point x:

$$S(x) = y_i + (x - x_i)[n_i + t(B + tA)],$$

where, as usual,

$$t = \frac{x - x_i}{x_{i+1} - x_i}.$$

The implementation of this algorithm substantially reduces the computing expenditure for the determination of the value $S(x)$.

1.1.5 Advice for users

The choice of the end conditions and the interpolation knots allows the properties of the interpolating spline to be controlled.

A. Choice of the end conditions

The choice of the end conditions is one of the central problems in function interpolation. This is of particular importance when a high level precision of the approximation of the given function $f(x)$ by the spline function $S(x)$ near the end points of the segment $[a, b]$ is required. The end values have a pronounced effect on the behavior of the spline $S(x)$ near points a and b, but as point x moves away from them, this effect becomes rapidly reduced.

The choice of the end conditions is often determined by the availability of additional information on behavior of the approximated function $f(x)$.

One can use the end conditions of the first type if the values of the first derivative $f'(x)$ at the end points of the segment $[a, b]$ are known.

Similarly, one can use the end conditions of the second type if the values of the second derivative $f''(x)$ at the end points of the segment $[a, b]$ are known.

If it is possible to choose between the end conditions of the first and second types, the preference should be given to the end conditions of the first type.

If $f(x)$ is a periodic function, the preference should be given to the end conditions of the third type.

If additional information is lacking, the so-called *natural end conditions* $S''(a) = 0$, $S''(b) = 0$ are frequently used. However, one must keep in mind that the precision of such approximation of the function $f(x)$ by the spline function $S(x)$ near the end points of the segment $[a, b]$ is greatly reduced. Sometimes the end conditions of the first and second types with different approximations of the corresponding derivatives are used. But the precision level of this approach is low.

The cumulative calculating experience suggests that in the case considered, the choice of the end conditions of the fourth type is most appropriate.

B. Choice of the interpolation knots

If the third derivative $f'''(x)$ of the function $f(x)$ is discontinuous at some points of the segment $[a, b]$, these points should be integrated in the number of the interpolation knots in order to increase the precision level of the approximation.

If the second derivative $f''(x)$ of the function $f(x)$ is discontinuous at some points of the segment $[a, b]$, then the spline oscillation near the break points may present a problem. In order to circumvent the problem, one can choose the interpolation knots so that every break point of the second derivative is an inner point of the partial segment $[x_i, x_{i+1}]$ with length h_i,

$$h_i = \alpha \min\{h_{i-1}, h_{i+1}\},$$

where $\alpha \ll 1$. Parameter α may be chosen by a computing experiment (often it is sufficient to put $\alpha = 0.01$).

There exists a number of recipes for minimizing the difficulties associated with the breaks of the first derivative $f'(x)$. Following is an example of one of the most simple recipes: decompose the approximation segment at the partial intervals where the derivative is continuous and construct a spline function on each of them.

1.1.6 Choice of the interpolating functions (pluses and minuses)

A. The Lagrange interpolating polynomial. For a given array

$$(x_i, y_i), \quad i = 0, 1, \dots, m-1, m$$

(Figure 3) *the Lagrange interpolating polynomial* is defined by formula

$$L_m(x) = \sum_{i=0}^{m} y_i \frac{\varphi_m(x)}{(x - x_i)\varphi'_m(x_i)},$$

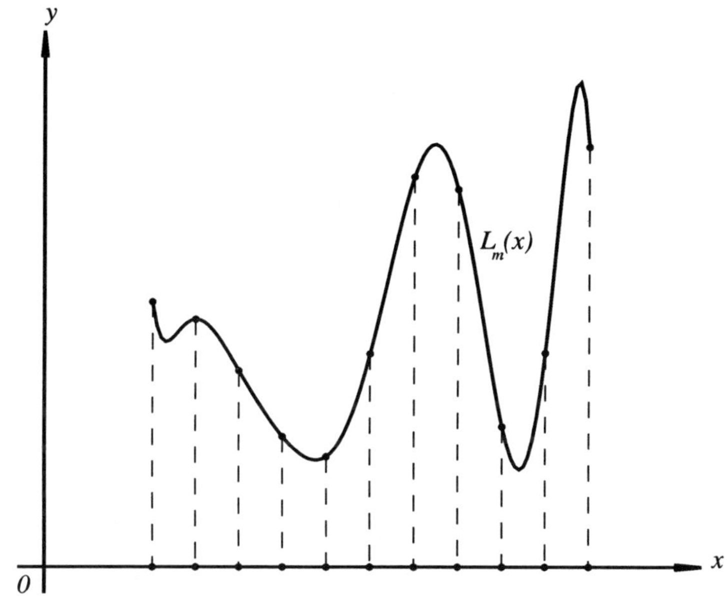

Figure 3.

where

$$\varphi_m(x) = \prod_{i=0}^{m} (x - x_i).$$

It makes sense to look at the properties of the Lagrange interpolating polynomial from two opposite sides and to discuss its main advantages and disadvantages individually.

Main advantages:

1) A graph of the Lagrange interpolating polynomial passes through every point of the array.

2) The constructed function is easy to describe (the number of the desired coefficients of the Lagrange interpolating polynomial on grid ω is equal to $m + 1$).

3) The constructed function has the continuous derivatives of every order.

4) The Lagrange interpolating polynomial is uniquely determined by the given array.

Main disadvantages:

1) A degree of the Lagrange interpolating polynomial depends on the number of the grid knots, and the more the number, the higher the degree of the Lagrange interpolating polynomial, and therefore, the greater the volume of calculations.

2) The change of one point in the given array causes the recalculation of all the coefficients of the Lagrange interpolating polynomial.

3) An addition of only one point to the given array raises the degree of the Lagrange interpolating polynomial and causes all the coefficients of the polynomial to be recalculated.

4) Under indefinite refinement of the grid the degree of the Lagrange interpolating polynomial increases indefinitely.

The behavior of the Lagrange interpolating polynomial under the indefinite grid refinement calls for particular attention.

Comment

On approximation of the continuous function by polynomial.

It is well known (Weierstrass, 1885) that every function continuous (all the more so smooth) on the segment may be approximated by the polynomial on the segment as closely as one likes.

Let us describe this fact in the language of formulas.

Let $f(x)$ be a function continuous on segment $[a, b]$. Then for every $\varepsilon > 0$ there exists a polynomial $P_n(x)$ such that for every x from the segment the following inequality

$$| P_n(x) - f(x) | < \varepsilon$$

is true (Figure 4). It should be noted that there exists an indefinitely large number of polynomials of the same degree approximating the function $f(x)$ with a given precision.

Let us construct a grid ω on the segment $[a, b]$. It is clear that, in general, its knots do not coincide with the intersection points of the graphs of polynomial $P_n(x)$ and function $f(x)$ (Figure 5). Therefore, the polynomial $P_n(x)$ is not an interpolating polynomial for the given grid.

On interpolation of the continuous function.

Under approximation of the continuous function by the Lagrange interpolating polynomial, the polynomial graph may deviate from the graph of the function as much as desired. Let us consider two examples.

Example 1 (Runge, 1901). Under an indefinite increase of the knot number, the following limit equality

$$\lim_{m \to \infty} \max_{-1 \le x \le 1} | f(x) - L_m(x) | = \infty$$

for function

$$f(x) = \frac{1}{1 + 25x^2}$$

on segment $[-1, 1]$ is true (Figure 6).

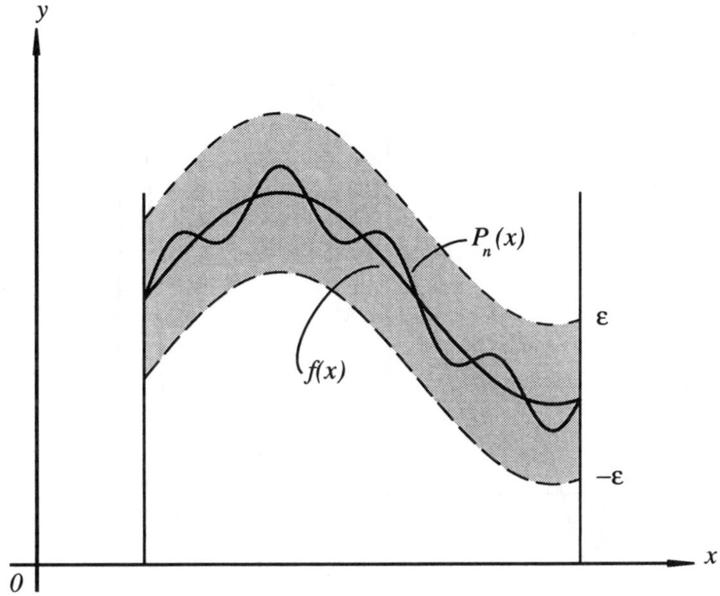

Figure 4.

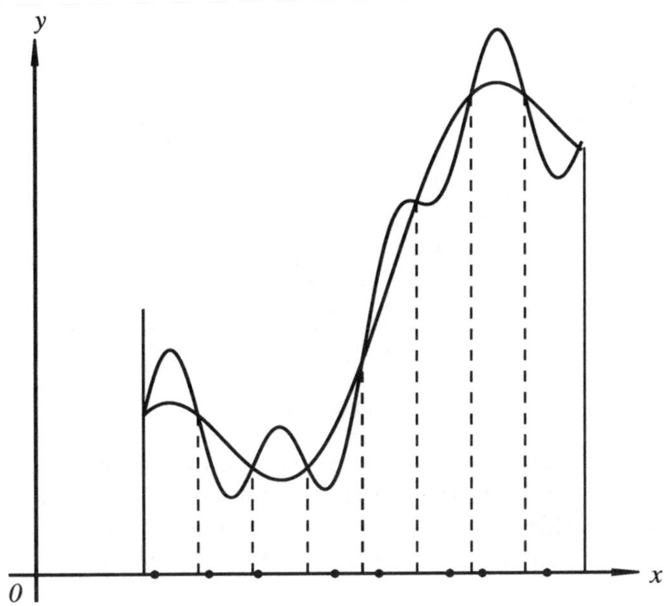

Figure 5.

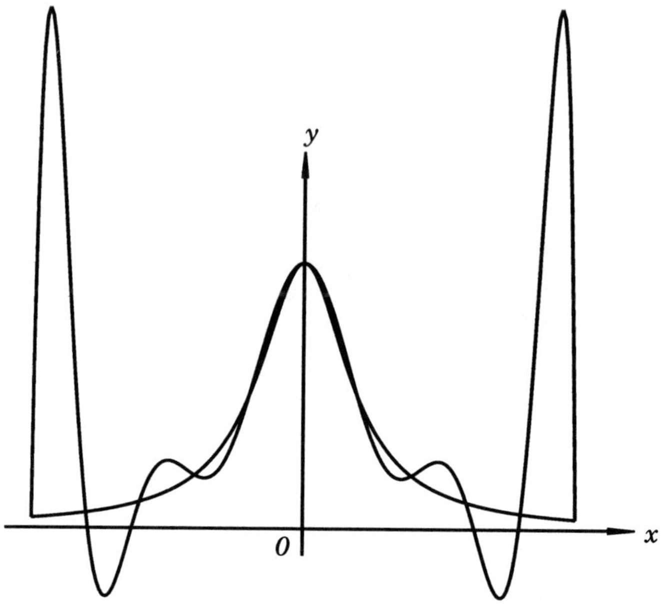

Figure 6.

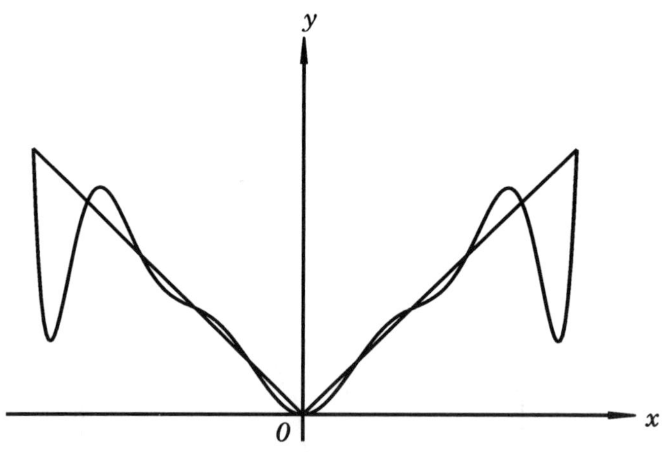

Figure 7.

Example 2 (Bernstein, 1912). A sequence of the Lagrange interpolating polynomials $\{L_m(x)\}$ constructed for the continuous function $f(x) = | x |$ on the uniform grids ω_m of segment $[-1, 1]$ *does not tend* to the function when the knot's number increases indefinitely (Figure 7).

B. Piecewise-linear interpolation. If the interpolating function is not smooth, the relationship between the advantages and disadvantages numbers may change substantially toward the advantages.

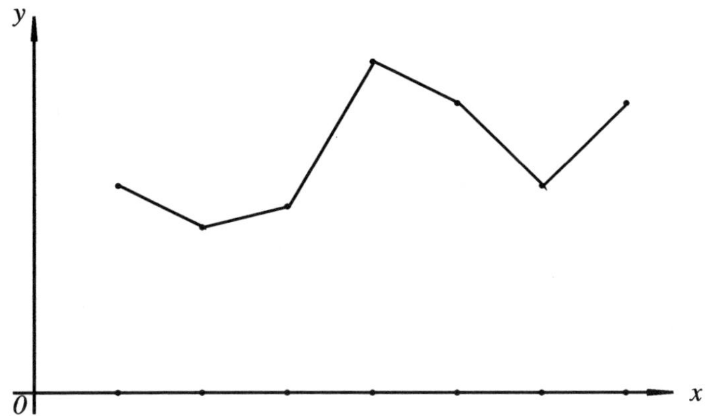

Figure 8.

Construct a piecewise linear function by sequentially joining the points (x_i, y_i) with the straight-line segments (Figure 8).

Main advantages:

1) A graph of the piecewise-linear function passes through every point of the given array.

2) The constructed function is easy to describe (the number of the desired coefficients of the corresponding linear functions for the grid ω is equal to $2m$).

3) The constructed function is uniquely determined by the given array.

4) A degree of the polynomials used for description of the interpolating function is independent of the knot's number (it is equal to 1).

5) A change of one point in the array causes four numbers (coefficients of two straight-line segments issued out of the new point) to be calculated.

6) An addition of only one point in the array requires four coefficients to be calculated.

Besides, a piecewise-linear function behaves sufficiently well under the grid refinement.

Main disadvantage:

Approximating piecewise-linear function is not differentiable: its first derivative has a break at the grid knots (*knots of interpolation*).

C. Spline interpolation. The approaches put forward above may be combined so that the number of the enumerated advantages of both approaches are retained under concurrent decrease of the disadvantage number. It may be obtained by the construction of the smooth interpolating spline function of degree p.

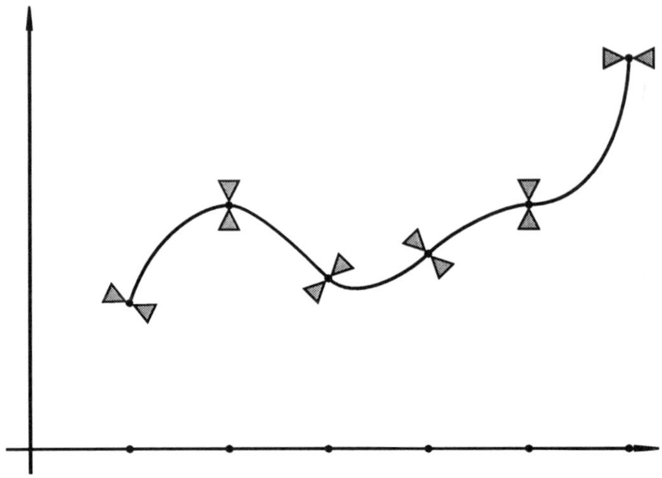

Figure 9.

Main advantages:

1) A graph of the constructed function passes through every point of the given array.

2) The constructed function is sufficiently easy to describe (the number of the desired coefficients of the corresponding polynomials for the grid ω is equal to $m(p+1)$).

3) The constructed function is uniquely determined by the given array.

4) A degree of the polynomials used for description of the interpolating function is independent of the knot's number and, consequently, does not change as the number increases.

5) The constructed function has the continuous derivatives up to and including $p-1$.

6) The constructed function possesses good approximating properties.

Brief reference. The name *"spline"* is not accidental: it is derived from an intimate connection between the smooth piecewise polynomial functions and the draftsman's scales (drawing-splines).

Let us consider a flexible ideally light-gauge bar passing through the support points of a planar array (x_i, y_i), $i = 0, 1, \ldots, m-1, m$ (Figure 9). According to the Bernoulli-Euler law, a linearized equation of the flexed bar has the following form:

$$EIS''(x) = -M(x),$$

where $S(x)$ is a bending, $M(x)$ — a bending moment *changed linearly* from support to support, and EI — a rigidity of the bar.

The function $S(x)$ describing the bar's shape is a polynomial of the third degree between any of two neighboring points (supports) of the given array and doubly continuously differentiable on the segment $[a, b]$.

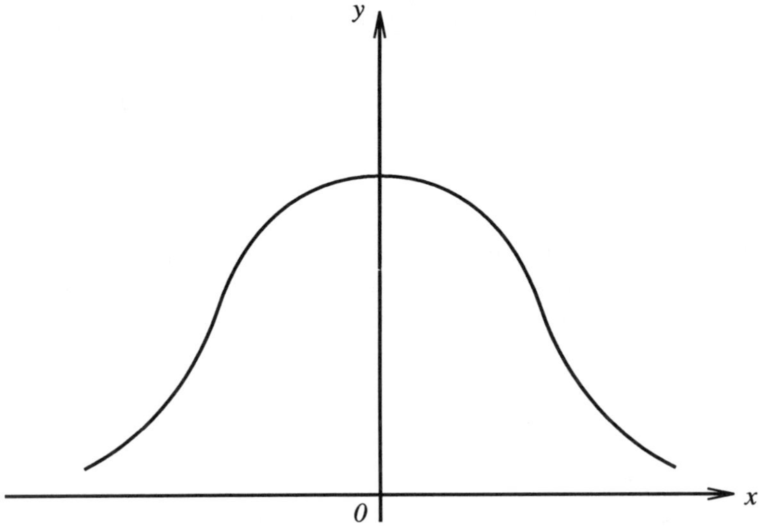

Figure 10.

Comment

On interpolation of a continuous function.

Unlike the Lagrange interpolating polynomials, a sequence of the interpolating cubic splines on a uniform grid tends to the interpolated function at all times. It is significant that a degree of the sequence convergence increases with the improved differentiated properties of the given function.

Example. The approximation error of function

$$f(x) = \frac{1}{1+25x^2}$$

by a cubic spline on the grid with the knot's number $m = 6$ is of the same order as the approximation error of the function by the Lagrange interpolating polynomial $L_5(x)$ on the same grid, and on the grid with the knot's number $m = 21$ the error is so small that there is no way to see it by the naked eye in the standard book drawing (Figure 10). (In this case the error of the corresponding interpolating polynomial $L_{20}(x)$ is near 10,000%.)

1.1.7 Properties of the interpolating cubic spline function

A. Approximate properties. The approximate properties of the interpolating spline depend on smoothness of a given function $f(x)$, the more the smoothness of the interpolated function, the more the approximation order and the more the convergence velocity under the grid refinement.

If the interpolated function $f(x)$ is continuous on segment $[a,b]$, that is, $f(x) \in C^0[a,b]$, then

$$\|f(x) - S(x)\|_C = \max_{x\in[a,b]} \mid f(x) - S(x) \mid \to 0$$

as

$$h = \max_{0\le i\le N-1} h_i \to 0.$$

If the interpolated function $f(x)$ has the first continuous derivative on the segment $[a,b]$, that is, $f(x) \in C'[a,b]$, and the interpolating spline $S(x)$ satisfies the end conditions of the first or third types, then

$$\|f(x) - S(x)\|_C = o(h), \quad \|f'(x) - S'(x)\|_C = o(1)$$

as $h \to 0$. In this case not only the spline tends to the interpolated function, but the first derivative of the spline tends to the first derivative of the function as well.

In the case where $f(x) \in C^4[a,b]$, the spline $S(x)$ approximates the function on the segment $[a,b]$ and its first and second derivatives approximate the functions $f'(x)$ and $f''(x)$, respectively,

$$\|f(x) - S(x)\|_C = o(h^4), \quad \|f'(x) - S'(x)\|_C = o(h^3),$$

$$\|f''(x) - S''(x)\|_C = o(h^2).$$

B. Extreme property. An interpolating cubic spline possesses one more useful property. Consider the following problem.

Problem. Construct a function $f(x)$ minimizing a functional

$$J(f) = \int_a^b (f''(x))^2 dx$$

on the class of functions from space $C^2[a,b]$, graphs of which pass through the points of array (x_i, y_i), $i = 0, 1, \ldots, m$.

Solution. Among all the functions of the space passing through the support points $(x_i, f(x_i))$, it is precisely the cubic spline $S(x)$ satisfying the end conditions

$$S''(a) = S''(b) = 0,$$

minimizing the functional $J(f)$.

Remark 1. Every so often the extreme property is used as a definition of the interpolating cubic spline.

Remark 2. It is interesting to note that the interpolating cubic spline possesses the extreme property described above on a more wide class of functions, namely, in space $W_2^2[a,b]$.

C. Construction of the interpolating spline curves with the help of the interpolating spline functions. Points of the arrays under consideration above are enumerated so that their abscissas generate a strongly

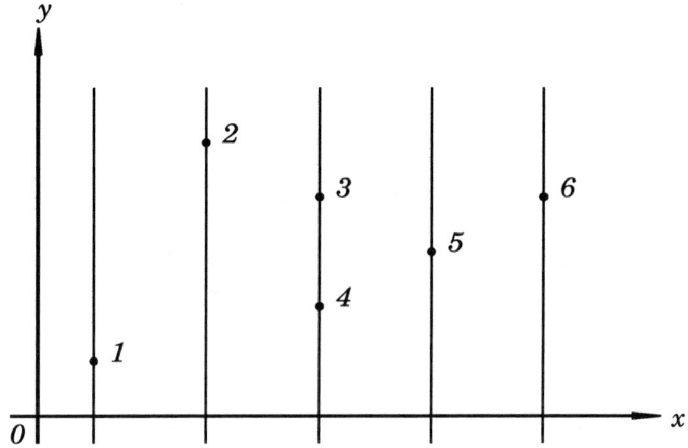

Figure 11.

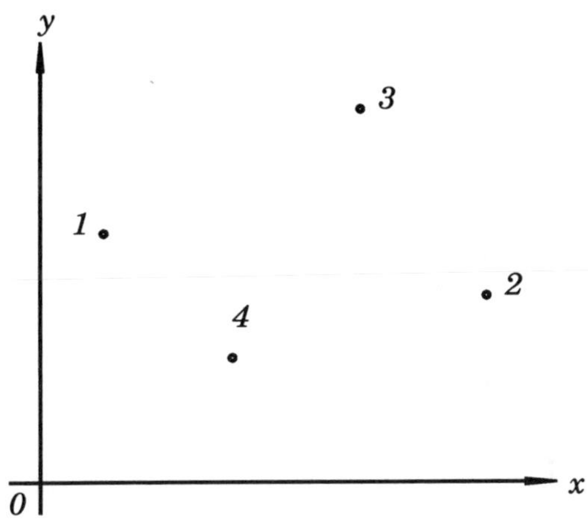

Figure 12.

increasing sequence. In particular, the case when various points of the array have the same abscissas (Figure 11) is not permissible. The fact defines both the choice of the approximated curves (graphs of functions) and the way of their construction.

At the same time, the method described above makes it possible to construct the interpolating curve sufficiently well in a more general case when the enumeration of the points of array

$$\mathbf{P} = \{\mathbf{P}_i(x_i, y_i), i = 0, 1, \dots, m\}$$

and their mutual arrangement on the plane, as a rule, are independent of one another (Figure 12).

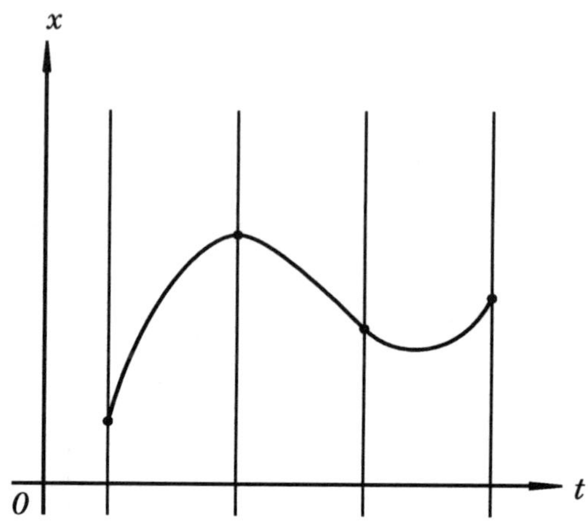

Figure 13a.

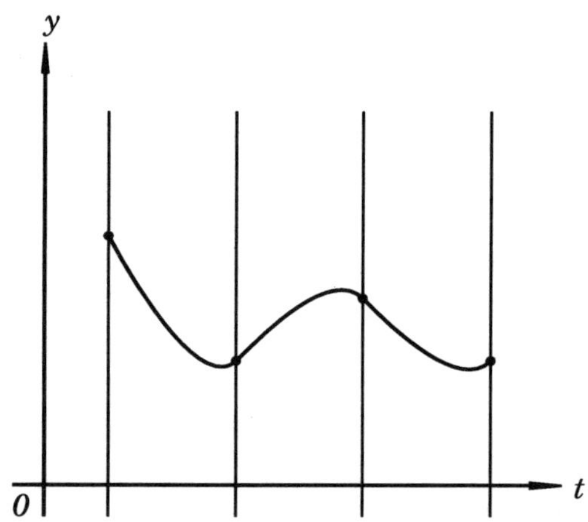

Figure 13b.

Moreover, one can consider the given array as nonplanar; that is,

$$\mathbf{P} = \{\mathbf{P}_i(x_i, y_i, z_i), i = 0, 1, \ldots, m\}.$$

It is clear that for the general problem to be solved it is necessary to extend the set of the admissible curves essentially by the addition of the closed curves, the curves with the self-intersection points, and the spatial curves. The curves from the set are conveniently described by the parametric equations

$$x = x(t), \quad y = y(t), \quad z = z(t), \tag{$*$}$$
$$\alpha \le t \le \beta.$$

The functions $x(t)$, $y(t)$, and $z(t)$ will be considered to be sufficiently smooth, for example, of class $C^1[\alpha, \beta]$ or of class $C^2[\alpha, \beta]$.

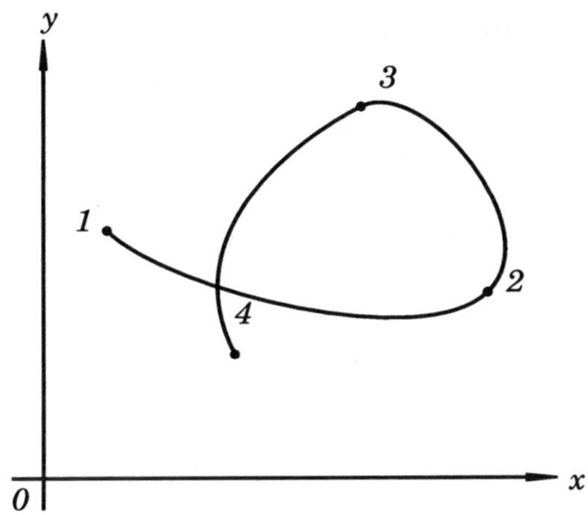

Figure 14.

To find the parametric equations of the curve passing sequentially through all points of the array, one can proceed as follows.

1st step. Take an arbitrary segment $[\alpha, \beta]$ as a domain of parameter t and construct an auxiliary grid

$$\alpha = t_0 < t_1 < \ldots < t_{m-1} < t_m = \beta$$

on the segment, where the knot's number coincides with the number of points in the array $\mathbf{P}$.

2nd step. For the given array $\mathbf{P}$ construct three new auxiliary arrays (two in the planar case (Figures 13a and 13b))

$$X = \{(t_i, x_i), i = 0, 1, \ldots, m\},$$

$$Y = \{(t_i, y_i), i = 0, 1, \ldots, m\},$$

$$Z = \{(t_i, z_i), i = 0, 1, \ldots, m\}.$$

3rd step. For each of the arrays X, Y, and Z, find the corresponding interpolating spline functions $x(t)$, $y(t)$, and $z(t)$.

As a result we have the parametric equations of the form $(*)$ for the curve passing through the points $\mathbf{P}_i$, $i = 0, 1, \ldots, m$ (see Figure 14 — the planar case).

Remark 1. The described approach allows one to construct the closed interpolating curves (for $\mathbf{P}_0 = \mathbf{P}_m$); to do this, the end conditions of the third type are required (when constructing the coordinate functions $x(t)$, $y(t)$, and $z(t)$).

Remark 2. The obtained curve is smooth but is not necessarily regular, because the opportunity of the equalities

$$x'(t^*) = 0, \quad y'(t^*) = 0, \quad z'(t^*) = 0$$

for some $t^* \in (\alpha, \beta)$ must not be ruled out. In addition, the curve may be with self-intersection points.

1.1.8 Implementation

The above algorithm for the construction of the interpolating cubic spline function and calculation of its values at an arbitrary point is realized in the program Spline (in Fortran). The routine call looks like

```
      call Spline(n,x,y,a,b,c,d,z,u,v,w,t,s,
     *            ind,ib,ax,bx,xx,sp,dsp,dsp2)
```

Input:
n — number of the interpolation knots
x — array of the interpolation knots (of size n)
y — array of the function values (of size n) at the interpolation knots
ax, bx — parameter values in the boundary conditions of the first and second types
xx — point where values of the spline and its first and second derivatives must be calculated
ind — mode type:
ind = 0 — the program calculates the spline coefficients
ind = 1 — the spline coefficients are known and the program calculates values of the spline and its first and second derivatives at the point xx
ib — index of the type of the boundary conditions:
ib = i (i=1-4) — the end conditions of the ith type
Output:
z — array of the spline coefficients (of size n)
sp — the spline value at point xx
dsp — value of the first derivative of the spline at point x
dsp2 — value of the second derivative of the spline at point x
Auxiliary variables:
a,b,c,d — work arrays (of size n)
s,t,u,v — work arrays (of size n+1)

Program Spline possesses a call of program Progon3 for solving the linear algebraic system with three-diagonal matrix (see Appendix A).

Before the program Spline call, it is necessary:

1) to describe the sizes of the arrays x,y,z,a,b,c,d,u,v keeping in mind that the size of the arrays u and v is higher by 1 than that of x, y, and others,

2) to assign the actual values to the elements of the arrays x and y and to the variables n, ib, and ind (ind = 0 before the first call and ind = 1 on the repeated calls).

Example. First construct a uniform grid with 21 knots x_i on segment $[0, 2\pi]$

$$x_i = \frac{\pi}{10} i, \quad i = 0, 1, \ldots, 20$$

and calculate the values of function $f(x) = \sin x$ at the knots. With these initial data, construct the interpolating cubic spline using the first type end conditions and then compare the values of the given function and the constructed spline as well as their first and second derivatives at points

$$\hat{x}_i = \frac{x_i + x_{i+1}}{2}, \quad i = 0, 1, \ldots, 19.$$

The program for the problem solution may be as follows:

```
      dimension x(81),y(81),a(82),b(82),c(82),d(82),
     *          z(82),u(82),v(82),w(82),s(82),t(82)
      data pi /3.1415926/
        n = 21
        h = 2.*pi/20
c
        do j = 1,21
          x(j) = h*(j-1)
          y(j) = sin(x(j))
        enddo
        ib = 1
        ax = 1.
        bx = 1.
      call spline (n,x,y,a,b,c,d,z,u,v,w,t,s,
     *             0,ib,ax,bx,xx,sp,dsp,d2sp)
        del0 = 0.
        del1 = 0.
        del2 = 0.
        do j = 1,41
          h  = 2.*pi/40.
          xx = h*(j-1)
          yy = sin(xx)
          d1 = cos(xx)
          d2 =-sin(xx)
        call spline (n,x,y,a,b,c,d,z,u,v,w,t,s,
     *               1,ib,ax,bx,xx,sp,dsp,d2sp)
        if (abs(yy-sp).gt.del0) del0= abs(yy-sp)
        if (abs(d1-dsp).gt.del1) del1= abs(d1-dsp)
        if (abs(d2-d2sp).gt.del2) del2= abs(d2-d2sp)
        enddo
        print *,' max|f(x)-S(x)|= ', del0
        print *,' max|f''(x)-S''(x)|= ', del1
        print *,' max|f"(x)-S"(x)|= ', del2
        stop
        end
```

The calculations, with the help of the program, furnish the following results:

$$\max |f(x) - S(x)| = 2.568960E - 05,$$
$$\max |f'(x) - S'(x)| = 6.669760E - 05,$$
$$\max |f''(x) - S''(x)| = 8.250833E - 03,$$

where maximum is computed on the set of points $\hat{x}_i$.

From the above relations it appears that the interpolating cubic spline in fact possesses good approximating properties.

The source of program `Spline` may be found on the applied diskette in `spline.for` file of subdirectory `INTERPOL` where the files containing the other examples using the program are also located. The detailed information on the file names and their contents is presented in `readme.txt` file of directory `SPLINES` and in Appendix B.

1.2 Smoothing cubic splines

1.2.1 Statement of the smoothing problem

Consider a grid ω and a set of numbers

$$y_0, y_1, \ldots, y_{m-1}, y_m.$$

Comment on the initial data. In applications, the case when values y_i in array

$$(x_i, y_i), \quad i = 0, 1, \ldots, m$$

are given with some error is frequent. In reality, this means that for every $i = 0, 1, \ldots, m$, an interval

$$(c_i, d_i) \quad \text{or} \quad (y_i - \delta_i, y_i + \delta_i)$$

is indicated and any number of the interval may be taken as y_i. Value y_i is convenient to interpret, for example, as the result of the measurement of some function $y(x)$ for the given value of variable x containing a random error. When solving the reconstruction problem of a function by the "experimental" data we see little reason for using the interpolation because the interpolating function will obediently reproduce any oscillations caused by the random component in array $\{y_i\}$. The approach based on the smoothing procedure capable of decreasing the element of randomness in the measurement results is more natural. Usually in such a problem it is required to find a function whose values for $x = x_i$, $i = 0, 1, \ldots, m$, lie in the corresponding intervals, and in addition the function obtains sufficiently good properties, for example, the continuous first and second derivatives, or a not-too-curvilinear graph.

A similar problem arises when it is necessary to construct a function by the precisely given array so that the function passes sufficiently smoothly near the given points. In other words, the desired function must smooth out the given array and must not interpolate it.

Let a grid ω and two sets of numbers be given as

$$y_0, y_1, \ldots, y_{m-1}, y_m \quad \text{and} \quad \Delta_0 > 0, \Delta_1 > 0, \ldots, \Delta_{m-1} > 0, \Delta_m > 0.$$

Problem. Construct a smooth (on the segment $[a,b]$) function $\sigma(x)$ the values of which, at the knots of grid ω, differ from numbers y_i by the given magnitudes $\Delta_i > 0$

$$\mid \sigma(x_i) - y_i \mid < \Delta_i, \quad i = 0, 1, \ldots, m-1, m.$$

Remark. The formulated smoothing problem consists in the reconstruction of a smooth function given by a table. It is clear that the problem has a set of various solutions. With the certain additional conditions on the desired function, the function may be constructed uniquely.

1.2.2 Definition of the smoothing cubic spline function

A function $S(x)$ defined on grid ω is called *a smoothing cubic spline function* if the function

1) is a cubic polynomial

$$S(x) = S_i(x) = a_0^{(i)} + a_1^{(i)}(x - x_i) + a_2^{(i)}(x - x_i)^2 + a_3^{(i)}(x - x_i)^3$$

on each partial segment $[x_i, x_{i+1}]$, $i = 0, 1, \ldots, m-1$,

2) has the continuous second derivatives on segment $[a,b]$, that is, the function is of class $C^2[a,b]$,

3) minimizes the functional

$$J(f) = \int_a^b (f''(x))^2 \, dx + \sum_{i=0}^{m} \frac{1}{\rho_i}(f(x_i - y_i)^2,$$

where y_i and $\rho_i > 0$ are given numbers, and

4) satisfies the boundary conditions of one of three types described below.

1.2.3 End (boundary) conditions

The end (boundary) conditions are given in the form of restrictions on the spline values and the values of the spline derivatives at the boundary knots of grid ω.

End (boundary) conditions of the first type

The values of the first derivative are given at the end points of the segment $[a,b]$

$$S'(a) = y'_0, \quad S'(b) = y'_m.$$

End (boundary) conditions of the second type

The values of the second derivative are given at the end points of segment $[a, b]$

$$S''(a) = 0, \quad S''(b) = 0.$$

End (boundary) conditions of the third type

are called *periodic*

$$S(a) = S(b), \quad S'(a) = S'(b), \quad S''(a) = S''(b).$$

Theorem. A cubic spline $S(x)$ minimizing the functional $(*)$ and satisfying the end conditions of one of the three types exists and is unique.

Remark 1. It is interesting to note that the smoothing cubic spline $S(x)$ minimizes the functional $J(f)$ in a more wide class of functions, namely, in space $W_2^2[a, b]$.

Definition. A cubic spline minimizing functional $J(f)$ and satisfying the end (boundary) conditions of the ith type is called *a smoothing cubic spline of the ith type.*

Remark 2. The spline function $S(x)$ is a cubic polynomial on each partial segment $[x_i, x_{i+1}]$, $i = 0, 1, \ldots, m-1$, and hence it is defined on the segment by four numbers. A total number of the partial segments is equal to m. Consequently, when we find $4m$ numbers

$$a_0^{(i)}, a_1^{(i)}, a_2^{(i)}, a_3^{(i)}, \quad i = 0, 1, \ldots, m-1$$

the desired spline function will be defined.

A condition $S(x) \in C^2[a, b]$ means a continuity of the function $S(x)$ and its derivatives $S'(x)$ and $S''(x)$ in all inner knots of grid ω. The knot's number is equal to $m-1$. Therefore, we have $3(m-1)$ conditions (equations) for the desired coefficients.

1.2.4 Construction of the smoothing cubic spline function

We describe a method of calculation where the number of the desired values is equal to $2m + 2$ instead of $4m$.

In each segment $[x_i, x_{i+1}]$, $i = 0, 1, \ldots, m-1$, the smoothing spline function is sought in the following form

$$S(x) = S_i(x) = z_i(1-t) + z_{i+1}t - \frac{h_i^2}{6}t(1-t)[(2-t)n_i + (1+t)n_{i+1}],$$

where

$$h_i = x_{i+1} - x_i, \quad t = \frac{x - x_i}{h_i},$$

and numbers z_i and n_i, $i = 0, 1, \ldots, m$, are a solution of the linear algebraic system, the form of which depends on the end condition type.

First, let us describe the search of numbers n_i.

For the *end conditions of the first and second types*, the system takes the form

$$a_0 n_0 + b_0 n_1 + c_0 n_2 = g_0,$$

$$b_0 n_0 + a_1 n_1 + b_1 n_2 + c_1 n_3 = g_1,$$

$$c_{i-2} n_{i-2} + b_{i-1} n_{i-1} + a_i n_i + b_i n_{i+1} + c_i n_{i+2} = g_i, \quad i = 2, 3, \ldots, m-2,$$

$$c_{m-3} n_{m-3} + b_{m-2} n_{m-2} + a_{m-1} n_{m-1} + b_{m-1} n_m = g_{m-1},$$

$$c_{m-2} n_{m-2} + b_{m-1} n_{m-1} + a_m n_m = g_m,$$

where

$$a_i = \frac{1}{3}(h_{i-1} + h_i) + \frac{1}{h_{i-1}^2}\rho_{i-1} + \left(\frac{1}{h_{i-1}} + \frac{1}{h_i}\right)^2 \rho_i + \frac{1}{h_i^2}\rho_{i+1},$$

$$i = 1, 2, \ldots, m-1,$$

$$b_i = \frac{1}{6}h_i - \frac{1}{h_i}\left(\left(\frac{1}{h_{i-1}} + \frac{1}{h_i}\right)\rho_i + \left(\frac{1}{h_i} + \frac{1}{h_{i+1}}\right)\rho_{i+1}\right), \qquad (*)$$

$$i = 1, 2, \ldots, m-2,$$

$$c_i = \frac{1}{h_i h_{i+1}}\rho_{i+1}, \quad i = 1, 2, \ldots, m-3,$$

$$g_i = \frac{y_{i+1} - y_i}{h_i} - \frac{y_i - y_{i-1}}{h_{i-1}}, \quad i = 1, 2, \ldots, m-1$$

($\rho_i \geq 0$ are the known numbers).

Coefficients

$$a_0, b_0, c_0, g_0, c_{m-2}, b_{m-1}, a_m, g_m$$

depend on the choice of the end conditions.

The end conditions of the first type:

$$a_0 = \frac{h_0}{3} + \frac{1}{h_0^2}(\rho_0 + \rho_1), \quad c_0 = \frac{1}{h_0 h_1}\rho_1,$$

$$b_0 = \frac{h_0}{6} - \frac{1}{h_0}\left(\frac{1}{h_0} + \frac{1}{h_1}\right)\rho_1 - \frac{1}{h_0^2}\rho_0), \quad g_0 = \frac{y_1 - y_0}{h_0} - y_0',$$

$$a_m = \frac{h_{m-1}}{3} + \frac{1}{h_{m-1}^2}(\rho_{m-1} + \rho_m), \quad c_{m-2} = \frac{1}{h_{m-2} h_{m-1}}\rho_{m-1},$$

$$b_{m-1} = \frac{h_{m-1}}{6} - \frac{1}{h_{m-1}}\left(\frac{1}{h_{m-1}} + \frac{1}{h_{m-2}}\right)\rho_{m-2} - \frac{1}{h_{m-1}^2}\rho_m),$$

$$g_m = y_m' - \frac{y_m - y_{m-1}}{h_{m-1}}.$$

The end conditions of the second type:

$$a_0 = 1, \quad b_0 = 1, \quad c_0 = 0, \quad g_0 = 0,$$

$$a_m = 1, \quad b_{m-1} = 1, \quad c_{m-2} = 0, \quad g_m = 0.$$

For the end conditions of the third type, the linear system for the desired numbers n_i, $i = 1, 2, \ldots, m$, may be written as follows:

$$c_{i-2}n_{i-2} + b_{i-1}n_{i-1} + a_i n_i + b_i n_{i+1} + c_i n_{i+2} = g_i, \quad i = 1, 2, \ldots, m.$$

All the coefficients are calculated by the formulas (*) (the magnitudes with indices k and $m + k$ are thought to be equal, that is, $n_0 = n_m$, $h_0 = h_m$, $a_0 = a_m$, etc.).

Important remark. The matrices of all three systems are nondegenerate, and, as a result, each of them has a unique solution.

When numbers n_i are found, the magnitudes z_i are easily determined by formulas

$$z_i = y_i - \rho_i D_i, \quad i = 1, 2, \ldots, m,$$

where

$$D_0 = \frac{1}{h_0}(n_1 - n_0), \quad D_m = -\frac{1}{h_{m-1}}(n_m - n_{m-1}),$$

$$D_i = \frac{1}{h_i}(n_{i+1} - n_i) - \frac{1}{h_{i-1}}(n_i - n_{i-1}), \quad i = 1, 2, \ldots, m - 1.$$

In the *periodic* case we have

$$h_m = h_0, \quad n_m = n_0, \quad n_1 = n_{m+1}$$

and

$$D_0 = D_m = \frac{1}{h_m}(n_1 - n_m) - \frac{1}{h_{m-1}}(n_m - n_{m-1}).$$

1.2.5 Choice of the weight coefficients

The choice of the *weight coefficients* ρ_i involved in functional (*) allows control of the behavior of the smoothing splines.

If all $\rho_i = 0$, then $z_i = y_i$ and the smoothing spline becomes interpolating. In particular, what this means is the more precise the given numbers y_i, the lesser must be the corresponding weight coefficients. If we want the spline to pass through point (x_k, y_k), then the corresponding weight factor ρ_k should be taken null.

The choice of magnitudes ρ_i is most important in the practical calculations.

Let Δ_i be a measurement error of value y_i. Then it is natural to require the smoothing spline $S(x)$ to satisfy the condition

$$| S(x_i) - y_i | \leq \Delta_i \tag{*}$$

or, what is the same — $\rho_i | D_i | \leq \Delta_i$.

In the simplest case the weight coefficients may be given, for example, by formula

$$\rho_i = c\Delta_i,$$

where c is some sufficiently small constant. But such a choice of weights ρ_i does not permit use of the "corridor" caused by the errors of magnitudes y_i. Below we describe a more rational and, at the same time, more tedious algorithm for the definition of the weight coefficients ρ_i.

If the values $D_i^{(k)}$ are found on the kth iteration, then we get

$$\rho_i^{(k+1)} = \begin{cases} \Delta_i / \mid D_i^{(k)} \mid & \text{for} \quad \mid D_i^{(k)} \mid > \varepsilon, \\ 0 & \text{for} \quad \mid D_i^{(k)} \mid \leq \varepsilon, \end{cases}$$

where ε is a small number chosen experimentally in view of the computer capacity, values Δ_i, and the precision of the solution of the linear algebraic system.

If the condition $(*)$ is broken on the kth iteration at the point x_i, then the last formula guarantees decreasing the corresponding weight coefficient ρ_i.

If

$$\mid S^{(k)}(x_i) - y_i \mid < \Delta_i,$$

then $\rho_i^{(k+1)} > \rho_i^{(k)}$ on the next iteration. Increasing ρ_i we can use the "corridor" $(*)$ more completely and ultimately construct a more smoothly changing spline.

1.2.6 Construction of the smoothing spline curves with the help of the smoothing spline functions

The points of the arrays under consideration above are enumerated so that their abscissas generate a strongly increasing sequence. In particular, the case when various points of the array have the same abscissas is not permissible. The fact defines both the choice of the approximated curves (graphs of functions) and the method of their construction.

At the same time, the method described above makes it possible to build the smoothing curve sufficiently well in a more general case when the enumeration of the points of array

$$\mathbf{P} = \{\mathbf{P}_i(x_i, y_i), i = 0, 1, \ldots, m\}$$

and their mutual arrangement on the plane are not connected. Moreover, one can consider the given array as nonplanar, that is,

$$\mathbf{P} = \{\mathbf{P}_i(x_i, y_i, z_i), i = 0, 1, \ldots, m\}.$$

It is clear that for the general problem to be solved, it is necessary to extend the admissible curves essentially by the addition of the closed curves, the self-intersecting curves, and the spartial curves. The curves from the set are conveniently described by the parametric equations of the following form:

$$x = x(t), \quad y = y(t), \quad z = z(t), \qquad (*)$$
$$\alpha \le t \le \beta.$$

In addition, let us require the functions $x(t)$, $y(t)$, and $z(t)$ to be sufficiently smooth, for example, of class $C^1[\alpha,\beta]$ or of class $C^2[\alpha,\beta]$.

To find the parametric equations of the smoothing curve, we proceed as follows.

1st step. Take an arbitrary segment $[\alpha,\beta]$ as a domain of parameter t and construct an auxiliary grid

$$\alpha = t_0 < t_1 < \ldots < t_{m-1} < t_m = \beta$$

on the segment, where the knot's number coincides with the point number in the array $\mathbf{P}$.

2nd step. For the given array $\mathbf{P}$ construct three new auxiliary arrays (two in the planar case)

$$\begin{aligned} \mathrm{X} &= \{(t_i, x_i), i = 0, 1, \ldots, m\}, \\ \mathrm{Y} &= \{(t_i, y_i), i = 0, 1, \ldots, m\}, \\ \mathrm{Z} &= \{(t_i, z_i), i = 0, 1, \ldots, m\}. \end{aligned}$$

3rd step. For each of the arrays X, Y, and Z, find the corresponding smoothing spline function $x(t)$, $y(t)$, and $z(t)$.

As a result, we have the parametric equations of the form $(*)$ for the curve smoothing the array $\mathbf{P}_i$, $i = 0, 1, \ldots, m$.

Remark 1. The described approach allows one to construct the closed smoothing curves (for $\mathbf{P}_0 = \mathbf{P}_m$); to construct the corresponding coordinate functions $x(t)$, $y(t)$, and $z(t)$, the end conditions of the third type are required.

Remark 2. The obtained curve is smooth but it is not necessarily regular, because the possibilities of the equalities

$$x'(t^*) = 0, \quad y'(t^*) = 0, \quad z'(t^*) = 0$$

for some $t^* \in (\alpha,\beta)$ must not be ruled out. In addition, the curve may have self-intersection points.

1.2.7 Implementation

The above algorithm for the construction of the smoothing cubic spline function and calculation of its values at an arbitrary point is realized in the program `Smspline` (in Fortran). The routine call looks like

```
      call Smspline(n,ib,ind,x,y,a,b,c,d,e,g,rho,ax,bx,
     *            u,v,w,p,q,r,s,t,aa,dd,zm,xx,sp,dsp,d2sp)
```

Input:
`n` — number of points in the array of the initial data
`x` — array containing x-coordinates of the initial points (of size `n`)
`y` — array containing y-coordinates of the initial points (of size `n`)
`rho` — array containing the values of the weight coefficients ρ_i (of size `n`)
`ind` — mode type:
`ind` = 0 — the program calculates the spline parameters
`ind` = 1 — the spline parameters are known and the program calculates the values of the spline and its first and second derivatives at point `xx`
`ax, bx` — parameters values in the boundary conditions of the first type
`xx` — point where the values of the spline and its first and second derivatives must be calculated
`ib` — type of the boundary conditions:
`ib = i (i=1-3)` — conditions of the `i`th type
Output:
`aa` — array (of size `n`), `aa(i)` contains the smoothing spline values at point `x(i)`
`z` — array (of size `n`), `z(i)` contains the values of the second derivatives of the smoothing spline at point `x(i)`
`sp` — value of spline at point `xx`
`dsp` — value of the first derivative of the spline at point `xx`
`d2sp` — value of the second derivative of the spline at point `xx`
Auxiliary variables:
`a,b,c,d,e,g,zm,aa,dd` — work arrays (of size `n`)
`u,v,w,p,q,r,s,t` — work arrays (of size `n+2`)

Program `Smspline` possesses a call of program `Progon5` for solving the linear algebraic system with a five-diagonal matrix (see Appendix A).

For illustration of the `Smspline` procedure call, consider the following example simulating the smoothing of the experiment data.

Suppose that file `input.dat` contains the results of 21 "measurements" of function $y = \sin x + \xi$ at the knots of grid

$$\omega = \{x_i = \frac{\pi}{20}(i-1), \quad i = 1, \ldots, 21\}.$$

(Here ξ is a random error of the measurement possessing the uniform distribution on segment [0.2, 0.2].)

Let us construct the smoothing spline satisfying the end conditions of the first type $S'(0) = 1$, $S'(\pi) = -1$ and tabulate the spline at 101 points with coordinates

$$x_j = \frac{\pi}{100}(j-1), \quad j = 1, 2, \ldots, 101.$$

Take magnitudes

$$\rho_i = 0.2, \quad i = 1, 2, \ldots, 21$$

as the weight factors.

The program for the problem solution may be as follows:

```
          real a(100),b(100),c(100),d(100),e(100),g(100)
          real w(100),u(100),v(100),s(100),zm(100)
          real r(100),t(100),p(100),q(100),rho(100)
          real aa(100),dd(100),x(100),y(100)
          data pi /3.1415926/
c
          open (1,file = 'input.dat')
c
          read (1,*) n
            do  i = 1,n
                read(1,*) x(i), y(i), rho(i)
            enddo
          read (1,*) ax, bx
c
          ib = 1
            call  Smspline(n,ib,0,x,y,a,b,c,d,e,g,rho,
     *                       ax,bx,u,v,w,p,q,r,s,t,aa,dd,
     *                       zm,xx,sp,dsp,d2sp)
          h = 2.*pi/100.
c
          do i = 1,101
               xx = h*(i-1)
               call  Smspline(n,ib,1,x,y,a,b,c,d,e,g,
     *                       rho,ax,bx,u,v,w,p,q,r,s,t,
     *                       aa,dd,zm,xx,sp,dsp,d2sp)
          write (2,1) xx,sp,sin(xx),dsp,cos(xx),d2sp,
     *                       -sin(xx)
          enddo
c
        stop
1       format(7(f8.4,2x))
        end
```

The results of calculations are shown in Figure 15 which also presents the graph of the interpolating spline satisfying the same end conditions and the sinusoid for comparison. The initial data are marked by dots.

From the Figure, one can see that the oscillations peculiar to the interpolating spline have nearly completely disappeared when we used the smoothing spline. The graphs of the first derivatives of the splines are

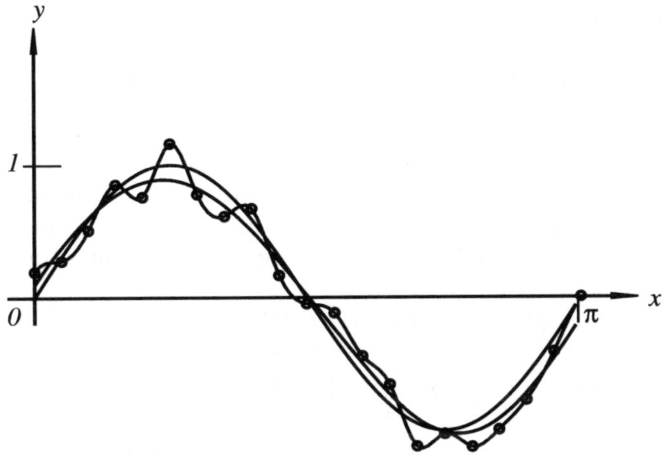

Figure 15.

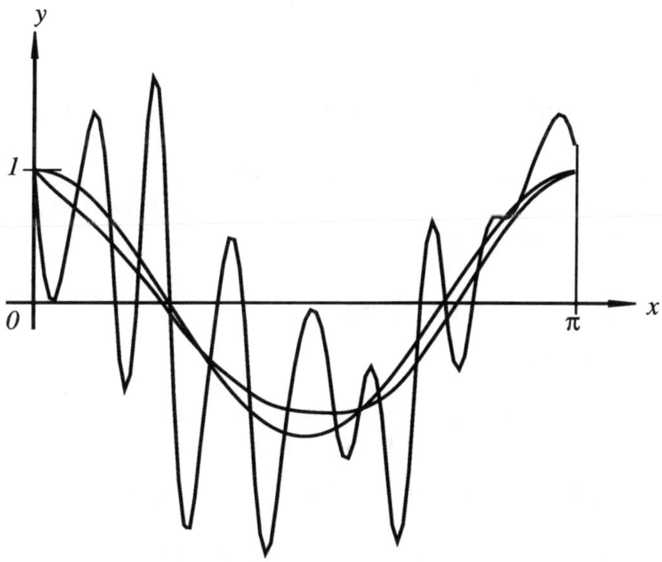

Figure 16.

shown in Figure 16; distinctions between the behavior of their second derivatives are more significant.

The source of program `Smspline` may be found on the applied diskette in `smspline.for` file of subdirectory `SMOOTH` where the files containing the other examples using the program are also located. The detailed information on the file names and their contents is presented in `readme.txt` file of directory `SPLINES` and in Appendix B.

1.3 Other types of splines

1.3.1 Linear space of cubic spline functions

A set of cubic spline functions constructed on segment $[a, b]$ by grid ω with $m + 1$ knots is a linear space of dimension $m + 3$:

1) A sum of two cubic spline functions constructed on segment $[a, b]$ by grid ω and a product of a cubic spline function constructed on segment $[a, b]$ by grid ω on an arbitrary number are also the cubic spline functions constructed by the grid.

2) Any cubic spline function constructed on segment $[a, b]$ by grid ω with $m + 1$ knots is defined uniquely by $m + 1$ values y_i at the knots and two end conditions, in all, by $m + 3$ parameters.

Choosing a basis in the linear space consisting of $m + 3$ linearly independent splines $\sigma_i(x)$, $i = 1, \ldots, m + 3$, we can write an arbitrary cubic spline function $\sigma(x)$ in the form of the linear combination

$$\sigma(x) = \sum_{i=1}^{m+3} \alpha_i \sigma_i(x)$$

with coefficients uniquely defined.

Remark. A similar description of the cubic spline functions is in wide use in the practice of calculations. The basis consisting of so-called *cubic B-spline functions (basic splines)* is especially convenient because the use of B-spline functions makes it possible to essentially lower the requirements on the memory capacity of the computer.

1.3.2 Cubic B-spline functions

A function of the form

$$B_i^{(0)}(x) = \begin{cases} 1, & x \in [x_i, x_{i+1}], \\ 0, & x \notin [x_i, x_{i+1}] \end{cases}$$

is called *a B-spline function of null degree* constructed on the number line by grid ω.

B-spline function of degree $k \geq 1$ constructed on the number line by grid ω is defined by the recurrent relation

$$B_i^{(k)}(x) = \frac{x - x_i}{x_{i+1} - x_i} B_i^{(k-1)}(x) + \frac{x_{i+k+1} - x}{x_{i+k+1} - x_{i+1}} B_{i+1}^{(k-1)}(x).$$

The graphs of the B-spline functions of the first $B_i^{(1)}(x)$ and the second $B_i^{(2)}(x)$ degrees are shown in Figures 17 and 18, respectively.

A B-spline function of an arbitrary degree k may differ from zero on the certain segment (defined by $k + 2$ knots) only.

It is more convenient to number the cubic B-spline functions in such a way that spline function $B_i^{(3)}(x)$ differs from zero on segment $[x_{i-2}, x_{i+2}]$.

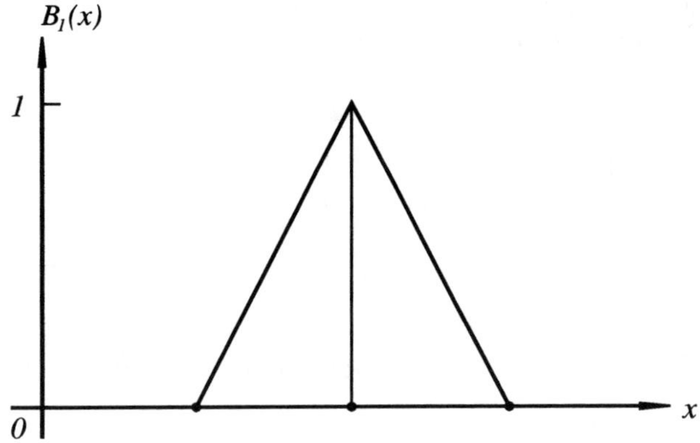

Figure 17.

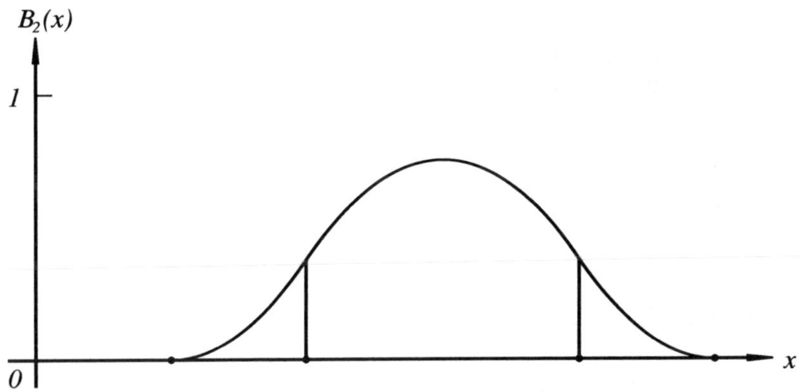

Figure 18.

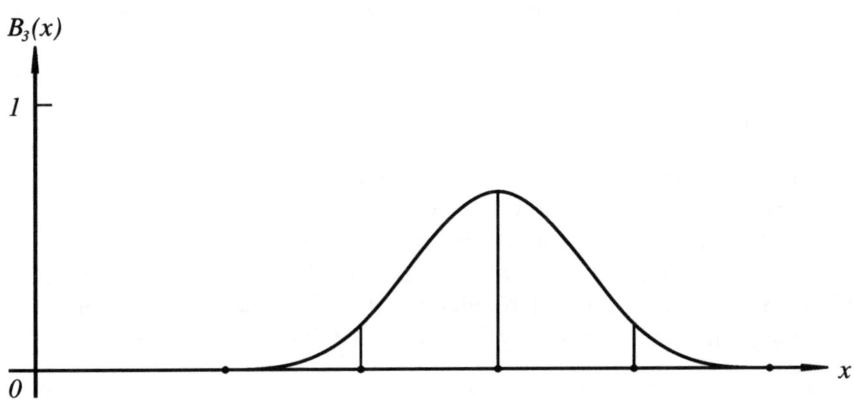

Figure 19.

Let us give a formula for a cubic B-spline function in the special case where the grid is uniform (with width h)

$$B_i^{(3)}(x) = \begin{cases} \frac{1}{6}\left(\frac{x-x_i}{h} - 2\right)^3, & x \in [x_i, x_{i+1}], \\ \frac{2}{3} - \frac{1}{2}\left(\left(\frac{x-x_i}{h}\right)^3 - \left(\frac{x-x_i}{h}\right)^2\right), & x \in [x_{i+1}, x_{i+2}], \\ \frac{2}{3} + \frac{1}{2}\left(\left(\frac{x-x_i}{h}\right)^3 - \left(\frac{x-x_i}{h}\right)^2\right), & x \in [x_{i+2}, x_{i+3}], \\ \frac{1}{6}\left(2 - \frac{x-x_i}{h}\right)^3, & x \in [x_{i+3}, x_{i+4}], \\ 0, & \text{otherwise.} \end{cases}$$

A typical graph of the cubic B-spline function is shown in Figure 19.

Function $B_i^{(3)}(x)$

a) is doubly differentiable on segment $[a, b]$, that is, the function is of class $C^2[a, b]$, and

b) differs from zero on four sequential segments $[x_{i-2}, x_{i-1}]$, $[x_{i-1}, x_i]$, $[x_i, x_{i+1}]$, and $[x_{i+1}, x_{i+2}]$.

The segment $[x_{i-2}, x_{i+2}]$ is called *a support* of function $B_i^{(3)}(x)$.

Let us add grid ω by arbitrary auxiliary knots

$$x_{-3} < x{-2} < x{-1} < a \quad \text{and} \quad b < x_{m+1} < x_{m+2} < x_{m+3}$$

and construct a set of $m + 3$ cubic B-splines

$$B_i^{(3)}(x), \quad i = -1, 0, \ldots, m, m+1$$

by the extended grid ω^*

$$x_{-3} < x{-2} < x{-1} < a < x_1 < \ldots < x_m < x_{m+1} < x_{m+2} < x_{m+3}.$$

The set is a basis of the linear space of the cubic spline functions on segment $[a, b]$. Thus any cubic spline function $S(x)$ constructed on segment $[a, b]$ by grid ω with $m + 1$ knots may be described in the following form:

$$S(x) = \sum_{i=-1}^{m+1} b_i B_i^{(3)}(x).$$

The coefficients b_i of the linear combinations are uniquely determined by conditions of the corresponding problem.

In case values y_i of an unknown function are given at the grid's knots and values y_0' and y_m' of its first derivative are given at the end points of the grid (the interpolation problem with the end (boundary) conditions of the first type), the coefficients may be calculated from the following system:

$$\begin{cases} b_{-1}B_{-1}'(x_0) + b_0 B_0'(x_0) + b_{-1}B_1'(x_0) = y_0', \\ b_{i-1}B_{i-1}'(x_i) + b_i B_i'(x_i) + b_{i+1}B_{i+1}'(x_i) = y_i, \quad i = 0, 1, \ldots, m, \\ b_{m-1}B_{m-1}'(x_m) + b_m B_m'(x_m) + b_{m+1}B_{m+1}'(x_m) = y_m'. \end{cases}$$

Eliminating the magnitudes b_{-1} and b_{m+1} we obtain the linear system with unknowns $b_0, \ldots, b_m$ and a three-diagonal matrix. The conditions (of the diagonal predominance)

$$\rho = \max_{|i-j|=1} \frac{h_i}{h_j} < \frac{1+\sqrt{3}}{2}$$

make it possible to use the sweep method.

Remark 1. The analogous linear systems arise in solving the other interpolation problems, too.

Remark 2. The use of the B-spline function in the interpolation problem allows essentially lower requirements on the memory capacity of computer in comparison to the algorithms described in Section 1.1 (true, in this case the number of operations increases).

Chapter 2

Spline functions of two variables

In a rectangle

$$R = [a,b] \times [c,d] = \{(x,y) \mid a \le x \le b, c \le y \le d\},$$

consider *a grid*

$$\omega = \omega_x \times \omega_y,$$

where

$$\begin{array}{ll} \omega_x: & a = x_0 < x_1 < \ldots < x_{m-1} < x_m = b, \\ \omega_y: & c = y_0 < y_1 < \ldots < y_{n-1} < y_n = d \end{array}$$

(Figure 1). Points

$$(x_i, y_j), \quad i = 1, \ldots, m-1, \quad j = 1, \ldots, n-1$$

are called *inner* knots of grid ω, points

$$(x_i, y_0), \quad (x_i, y_n), \quad i = 1, \ldots, m-1,$$

$$(x_0, y_j), \quad (x_m, y_j), \quad j = 1, \ldots, n-1$$

— *boundary* knots of the given grid, and points

$$(x_0, y_0), \quad (x_0, y_n), \quad (x_m, y_0), \quad (x_m, y_n)$$

— its *corner* knots.

Grid ω divides the rectangle R into cells

$$R_{ij} = \{(x,y) \mid x_i \le x \le x_{i+1}, y_j \le y \le y_{j+1}\},$$

$$i = 0, 1, \ldots, m-1, \quad j = 0, 1, \ldots, n-1$$

(Figure 2).

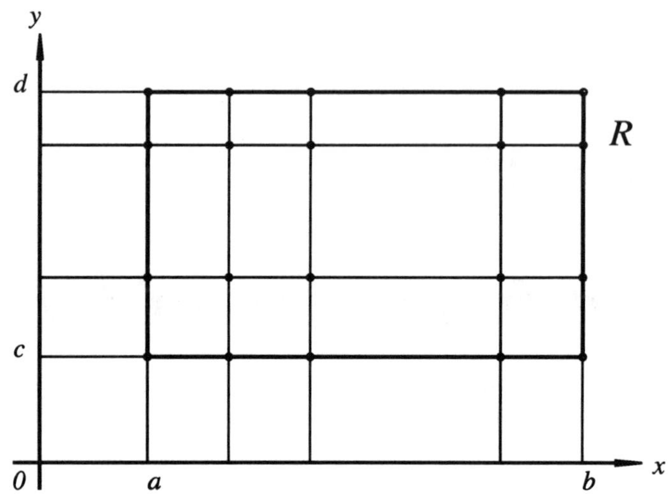

Figure 1.

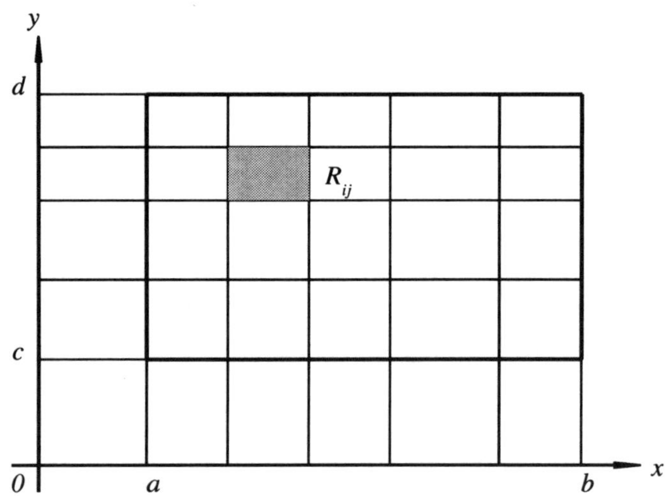

Figure 2.

Function $S(x,y)$ given on the rectangle R is called *a bicubic spline*, or *a bicubic spline function* if

1) the function is a polynomial of the third order with respect to variable x as well as to variable y in every rectangular cell R_{ij}, that is,

$$S(x,y) = \sum_{k=0}^{3}\sum_{l=0}^{3} a_{kl}^{(i,j)}(x-x_i)^k(y-y_j)^l,$$

$$i = 0,1,\ldots,m-1, \quad j = 0,1,\ldots,n-1;$$

2) the function $S(x, y)$ and its derivatives

$$\frac{\partial S}{\partial x}, \quad \frac{\partial S}{\partial y}, \quad \frac{\partial^2 S}{\partial x^2}, \quad \frac{\partial^2 S}{\partial x \partial y}, \quad \frac{\partial^2 S}{\partial y^2}, \quad \frac{\partial^3 S}{\partial x^2 \partial y}, \quad \frac{\partial^3 S}{\partial x \partial y^2}, \quad \frac{\partial^4 S}{\partial x^2 \partial y^2}$$

are continuous in the rectangle R, that is, $S(x, y) \in C^{2,2}(R)$.

Remark 1. Indices i and j of coefficient $a_{kl}^{(i,j)}$ indicate that the system of numbers defining spline function $S(x, y)$ is of its own for every cell R_{ij}.

Remark 2. In every rectangular cell R_{ij} the bicubic spline $S(x, y)$ may be written in the form

$$S(x, y) = \sum_{l=0}^{3} \left(\sum_{k=0}^{3} a_{kl}^{(i,j)} (x - x_i)^k \right) (y - y_j)^l$$

or in the form

$$S(x, y) = \sum_{k=0}^{3} \left(\sum_{l=0}^{3} a_{kl}^{(i,j)} (y - y_j)^l \right) (x - x_i)^k.$$

The first record allows one to say that for every cell R_{ij} the bicubic spline $S(x, y)$ is a cubic polynomial with respect to variable $y - y_j$ coefficients of which are the cubic polynomials with respect to variable $x - x_i$. The second record has a similar meaning.

This fact gives grounds to expect that a construction scheme of the two-dimensional bicubic splines leans substantially upon the algorithm of constructing the one-dimensional cubic splines.

In every rectangular cell R_{ij}, the spline function $S(x, y)$ is defined by sixteen coefficients:

$$\begin{array}{cccc} a_{00}^{(i,j)} & a_{10}^{(i,j)} & a_{20}^{(i,j)} & a_{30}^{(i,j)} \\ a_{01}^{(i,j)} & a_{11}^{(i,j)} & a_{21}^{(i,j)} & a_{31}^{(i,j)} \\ a_{02}^{(i,j)} & a_{12}^{(i,j)} & a_{22}^{(i,j)} & a_{32}^{(i,j)} \\ a_{03}^{(i,j)} & a_{13}^{(i,j)} & a_{23}^{(i,j)} & a_{33}^{(i,j)}. \end{array}$$

The total number of the rectangular cells is mn; that is, for a complete description of a bicubic spline on the grid ω, it is necessary to find $16mn$ numbers

$$a_{pq}^{(i,j)}, \qquad p = 0, 1, 2, 3, \quad q = 0, 1, 2, 3,$$
$$i = 0, 1, \ldots, m - 1, \quad j = 0, 1, \ldots, n - 1.$$

The problems most often considered are the interpolation and smoothing problems. In these cases it is required to build one or the other spline by a given array of points

$$(x_{ij}, y_{ij}, z_{ij}), \quad i = 1, \ldots, m, \quad j = 1, \ldots, n$$

(Figure 3).

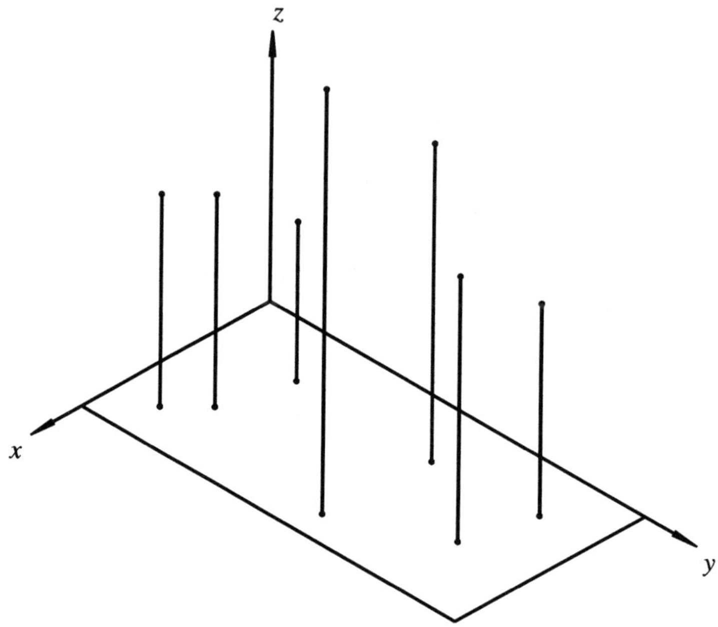

Figure 3.

In the *interpolation problem*, it is necessary that the spline graph pass through the points

$$(x_{ij}, y_{ij}, z_{ij}), \quad i = 0, 1, \ldots, m, \quad j = 0, 1, \ldots, n.$$

The requirement imposes a sequence of additional conditions (equations) on its coefficients. But these conditions fall short of a unique construction of the spline. A set of the needed additional conditions are usually given as the values of its lowest derivatives at the boundary knots of the grid ω — *boundary conditions.* The possibility of the choice of the boundary conditions permits one to construct splines with various properties.

In the *smoothing problem*, the spline is constructed in such a way that its graph goes nearly to points

$$(x_{ij}, y_{ij}, z_{ij}), \quad i = 0, 1, \ldots, m, \quad j = 0, 1, \ldots, n,$$

rather than through them. A measure of the nearness may be defined differently, resulting in a wide variety of smoothing spline types.

In this chapter we dwell on the spline functions of two variables which are the direct extension of the corresponding one-dimensional splines. But it should be noted that the considered two-dimensional splines don't exhaust their variety which is defined by the choice of the boundary conditions, the given array of points, the class of functions used for construction of the elementary fragments, etc.

2.1 Interpolating bicubic splines

2.1.1 Statement of the interpolating problem

Let a grid ω

$$a = x_0 < x_1 < \ldots < x_{m-1} < x_m = b,$$

$$c = y_0 < y_1 < \ldots < y_{n-1} < y_n = d$$

be given in rectangle $R = [a, b] \times [c, d]$. Consider a set of numbers

$$z_{ij}, \qquad i = 0, 1, \ldots, m, \quad j = 0, 1, \ldots, n.$$

Problem. Find a smooth (on the rectangle $R = [a, b] \times [c, d]$) function $\sigma(x, y)$ the values of which coincide with given numbers at the knots of the grid ω, that is

$$\sigma(x_i, y_j) = z_{ij}, \quad i = 0, 1, \ldots, m, \quad j = 0, 1, \ldots, n.$$

Remark. The formulated problem consists of the reconstruction of a smooth function given by a table. It is clear that the problem has a set of various solutions. With certain additional conditions for the desired function, the function may be uniquely constructed.

2.1.2 Definition of the interpolating bicubic spline function

A function $S(x, y)$ defined on the grid ω is called *an interpolating bicubic spline function* if the function

1) is a bicubic polynomial

$$S(x, y) = \sum_{p=0}^{3} \sum_{q=0}^{3} a_{p,q}^{(i,j)} (x - x_i)^p (y - y_j)^q$$

in each cell

$$R_{ij} = \{(x, y) \mid x_i \le x \le x_{i+1}, \quad y_j \le y \le y_{j+i}\},$$

$$i = 0, 1, \ldots, m, \quad j = 0, 1, \ldots, n$$

2) is a function of class $C^{2,2}(R)$, and

3) satisfies the conditions

$$S(x_i, y_j) = z_{ij}, \quad i = 0, 1, \ldots, m, \quad j = 0, 1, \ldots, n \qquad (*)$$

To construct the spline it is necessary to find $16mn$ coefficients $a_{p,q}^{(i,j)}$. The relations $(*)$ give $(m + 1)(n + 1)$ conditions. In concert with the requirement $S(x, y) \in C^{2,2}(R)$ we have $16mn - 2(m + n) - 8$ conditions (equations) on coefficients $a_{p,q}^{(i,j)}$.

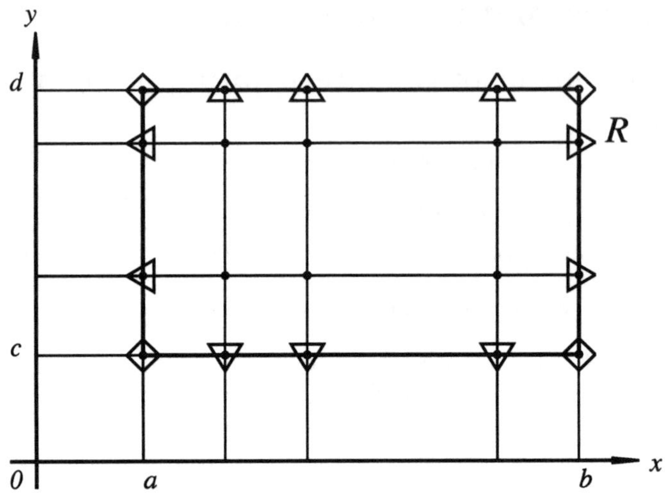

Figure 4.

2.1.3 Boundary conditions

The deficient $2(m+n+4)$ conditions are given in the form of restrictions on the values of the spline derivatives at the boundary and corner knots of the grid ω. In the construction of the interpolating bicubic spline the boundary conditions of four following types are most often used.

Boundary conditions of the first type

The values of the first partial derivatives of the desired function with respect to x and y are given at the boundary knots of the grid ω and the values of the second mixed derivative S_{xy} — at the corner knots:

$$\begin{array}{lll} \frac{\partial S}{\partial x}(x_i, y_j) = z_{ij}^{(x)}, & i = 0, m, & j = 0, 1, \ldots, n, \\ \frac{\partial S}{\partial y}(x_i, y_j) = z_{ij}^{(y)}, & i = 0, 1, \ldots, m, & j = 0, n, \\ \frac{\partial^2 S}{\partial x \partial y}(x_i, y_j) = z_{ij}^{(xy)}, & i = 0, m, & j = 0, n. \end{array}$$

The total number of the first type boundary conditions is equal to $2m + 2n + 8$.

A way of the spline definition by the boundary conditions of the first type is illustrated in Figure 4. The knots where the values of the spline S are given are marked off by the bullets $\bullet$, the knots where the values of the first partial derivatives S_x and S_y are given are marked off by horizontal $\triangleleft$, $\triangleright$ and vertical $\triangle$, $\triangledown$ arrows, respectively, and the knots where the values of the mixed derivative S_{xy} are given are marked off by rhombus $\diamondsuit$.

Boundary conditions of the second type

The values of the second partial derivatives of the desired function with respect to x and y are given at the boundary knots of the grid ω and the values of the mixed derivative S_{xxyy} — at the corner knots:

$$\begin{array}{lll} \frac{\partial^2 S}{\partial x^2}(x_i, y_j) = z_{ij}^{(x)}, & i = 0, m, & j = 0, 1, \ldots, n, \\ \frac{\partial^2 S}{\partial y^2}(x_i, y_j) = z_{ij}^{(y)}, & i = 0, 1, \ldots, m, & j = 0, n, \\ \frac{\partial^4 S}{\partial x^2 \partial y^2}(x_i, y_j) = z_{ij}^{(xy)}, & i = 0, m, & j = 0, n. \end{array}$$

The total number of the second type boundary conditions is equal to $2m + 2n + 8$.

A way of the spline definition by the boundary conditions of the second type may be also illustrated in Figure 4. The knots where the values of the spline S are given are marked off by the bullets $\bullet$, the knots where the values of the second partial derivatives S_{xx} and S_{yy} are given are marked off by horizontal $\triangleleft$, $\triangleright$ and vertical $\triangle$, $\bigtriangledown$ arrows, respectively, and the knots where the values of the mixed derivative S_{xxyy} are given are marked off by rhombus $\diamondsuit$.

Boundary conditions of the third type

are called *periodic*. In this case the spline $S(x, y)$ and its partial derivatives must be doubly periodic; their period with respect to variable x is equal to $T_x = b - a$ and with respect to variable y is equal to $T_y = d - c$:

$$\begin{array}{lll} S(x_0, y_j) = S(x_m, y_j), & j = 0, 1, \ldots, n & \\ S(x_i, y_0) = S(x_i, y_n), & i = 0, 1, \ldots, m & \\ \frac{\partial^k S}{\partial x^k}(x_0, y_j) = \frac{\partial^k S}{\partial x^k}(x_m, y_j), & j = 0, 1, \ldots, n, & k = 1, 2, \\ \frac{\partial^l S}{\partial y^l}(x_i, y_0) = \frac{\partial^l S}{\partial y^l}(x_i, y_n), & i = 0, 1, \ldots, m, & l = 1, 2, \\ \frac{\partial^{2k} S}{\partial x^k \partial y^k}(x_0, y_j) = \frac{\partial^{2k} S}{\partial x^k \partial y^k}(x_m, y_j), & j = 0, 1, \ldots, n, & k = 1, 2, \\ \frac{\partial^{2k} S}{\partial x^k \partial y^k}(x_i, y_0) = \frac{\partial^{2k} S}{\partial x^k \partial y^k}(x_i, y_n), & i = 0, 1, \ldots, m, & k = 1, 2. \end{array}$$

The conditions on spline function S are among the conditions $(*)$, and the total number of the conditions on its derivatives is equal to $2(2m+2n+4)$.

Boundary conditions of the fourth type

require that the derivatives of the desired spline

$$\frac{\partial S}{\partial x}, \quad \frac{\partial S}{\partial y}, \quad \frac{\partial^2 S}{\partial x^2}, \quad \frac{\partial^2 S}{\partial x \partial y}, \quad \frac{\partial^2 S}{\partial y^2},$$

$$\frac{\partial^3 S}{\partial x^3}, \quad \frac{\partial^3 S}{\partial x^2 \partial y}, \quad \frac{\partial^3 S}{\partial x \partial y^2}, \quad \frac{\partial^4 S}{\partial x^3 \partial y}, \quad \frac{\partial^4 S}{\partial x^2 \partial y^2}, \quad \frac{\partial^5 S}{\partial x^3 \partial y^2}$$

be continuous on the lines $x = x_1$ and $x = x_{m-1}$;

and the derivatives

$$\frac{\partial S}{\partial x}, \quad \frac{\partial S}{\partial y}, \quad \frac{\partial^2 S}{\partial x^2}, \quad \frac{\partial^2 S}{\partial x \partial y}, \quad \frac{\partial^2 S}{\partial y^2},$$

$$\frac{\partial^3 S}{\partial x^2 \partial y}, \quad \frac{\partial^3 S}{\partial x \partial y^2}, \quad \frac{\partial^3 S}{\partial y^3}, \quad \frac{\partial^4 S}{\partial x^2 \partial y^2}, \quad \frac{\partial^4 S}{\partial x \partial y^3}, \quad \frac{\partial^5 S}{\partial x^2 \partial y^3}$$

be continuous on lines $y = y_1$ and $y = y_{n-1}$.

Comment. The spline satisfied the fourth type boundary conditions has the smoothness higher than usual: its derivatives

$$\frac{\partial^p S}{\partial x^q \partial y^r}, \quad 1 \le p \le 6, \quad 0 \le q \le 3, \quad 0 \le r \le 3$$

are continuous in the rectangles

$$[x_k, x_{k+2}] \times [y_l, y_{l+2}], \quad k = 0, m-2, \quad l = 0, n-2$$

adjacent to the vertices of the rectangle R.

Remark. In addition to the described conditions there exist mixed boundary conditions, that is, the conditions of the distinct types for the distinct variables. In this case it is necessary to coordinate these conditions: if the first type conditions are given with respect to variable x, and the second type conditions with respect to variable y, then at the vertices of the rectangle R we need to take the partial derivatives S_{xyy}, etc.

2.1.4 Construction of the interpolating bicubic spline function

To describe the algorithm, the given information is conveniently written in the form of the following table:

Table 1

$z_{0,n}^{(xy)}$	$z_{0,n}^{(y)}$	$z_{1,n}^{(y)}$	$\dots$	$z_{m-1,n}^{(y)}$	$z_{m,n}^{(y)}$	$z_{m,n}^{(xy)}$
$z_{0,n}^{(x)}$	$z_{0,n}$	$z_{1,n}$	$\dots$	$z_{m-1,n}$	$z_{m,n}$	$z_{m,n}^{(x)}$
$z_{0,n-1}^{(x)}$	$z_{0,n-1}$	$z_{1,n-1}$	$\dots$	$z_{m-1,n-1}$	$z_{m,n-1}$	$z_{m,n-1}^{(x)}$
$\dots$	$\dots$		$\dots$		$\dots$	$\dots$
$z_{0,1}^{(x)}$	$z_{0,1}$	$z_{1,1}$	$\dots$	$z_{m-1,1}$	$z_{m,1}$	$z_{m,1}^{(x)}$
$z_{0,0}^{(x)}$	$z_{0,0}$	$z_{1,0}$	$\dots$	$z_{m-1,0}$	$z_{m,0}$	$z_{m,0}^{(x)}$
$z_{0,0}^{(xy)}$	$z_{0,0}^{(y)}$	$z_{1,0}^{(y)}$	$\dots$	$z_{m-1,0}^{(y)}$	$z_{m,0}^{(y)}$	$z_{m,0}^{(xy)}$

The values z_{ij} of the spline in the knots of the grid ω are arranged in the interior part of the table. The bounding rows and columns are the numerical values of the corresponding derivatives (for the boundary conditions of the first and second types only).

The algorithm of the construction of the interpolating bicubic spline is based on the fact that, with the fixed value of one of the variables (for example, y), both the spline and its partial derivative with respect to y are the interpolating cubic splines with respect to variable x.

1st step of the algorithm

With every row of the table 1 including the bounding rows (when present) we construct the one-dimensional interpolating cubic spline with respect to variable x using the end conditions taken from the bounding columns (when present). The construction of every one-dimensional spline may be reduced to the determination of numbers

$$\mu_{ij}^{(x)}, \quad i = 0, 1, \ldots, m \quad j = 0, 1, \ldots, n$$

which coincide with the values of the derivative S_x of spline S at the knots of grid ω. These numbers are conveniently written (without the bounding columns) in the form of the following table:

Table 2

$\mu_{0,n}^{(xy)}$	$\mu_{1,n}^{(xy)}$	...	$\mu_{m,n}^{(xy)}$
$\mu_{0,n}^{(x)}$	$\mu_{1,n}^{(x)}$	...	$\mu_{m,n}^{(x)}$
...	...	...	...
$\mu_{0,0}^{(x)}$	$\mu_{1,0}^{(x)}$	...	$\mu_{m,0}^{(x)}$
$\mu_{0,0}^{(xy)}$	$\mu_{1,0}^{(xy)}$	...	$\mu_{m,0}^{(xy)}$

For the boundary conditions of the first and second types, the values of the derivatives S_{xy} and S_{xyy} at the boundary knots of grid ω are sited in the upper and lower rows of the table, respectively. For the boundary conditions of the third and fourth types with respect to variable y, the bounding rows are absent.

2nd step of the algorithm

In the ith column of table 2 we construct the interpolating cubic spline $\tilde{S}$ being the partial derivative of the desired spline S with respect to variable x on line $x = x_i$

$$\tilde{S}(x_i, y) = \frac{\partial S}{\partial x}(x_i, y),$$

$$i = 0, 1, \ldots, m.$$

The construction of every spline $S(x_i, y)$ is reduced to the determination of numbers $\mu_{ij}^{(xy)}$ which are equal to the values of the derivative of the auxiliary spline $\tilde{S}$ with respect to variable y and thus to the values of derivative S_{xy} of the desired spline at the knots of grid ω.

3rd step of the algorithm

The third step is similar to the 1st step: by the data of table 1 we construct the one-dimensional interpolating splines $S(x_i, y)$, $i = 0, 1, \ldots, m$, with respect to variable y satisfying the boundary conditions of the corresponding type. As a result, we find the numbers

$$\mu_{ij}^{(y)} = \frac{\partial S}{\partial y}(x_i, y_j), \quad i = 0, 1, \ldots, m, \quad j = 0, 1, \ldots, n.$$

Upon completion of the 3rd step, the following magnitudes

$$z_{ij}, \quad \mu_{ij}^{(x)}, \quad \mu_{ij}^{(y)}, \quad \mu_{ij}^{(xy)}$$

will be known at every knot of grid ω. The numbers completely determine the interpolating bicubic spline satisfying the given boundary conditions.

Algorithm for calculation of the spline values

To calculate a value of the interpolating bicubic spline at arbitrary point (x, y) of rectangle R, let us define the following parameters:

number i such that $x \in [x_i, x_{i+1}]$,

number j such that $y \in [y_j, y_{j+1}]$,

steps $h_i = x_{i+1} - x_i$ and $l_j = y_{j+1} - y_j$ of grid ω with respect to x and y, respectively, and

numbers $t = (x - x_i)/h_i$ and $u = (y - y_j)/l_j$.

Introduce four auxiliary functions

$$\begin{aligned} \varphi_1(\xi) &= (1-\xi)^2(1+2\xi), & \varphi_2(\xi) &= \xi^2(3-2\xi), \\ \varphi_3(\xi) &= \xi(1-\xi)^2, & \varphi_4(\xi) &= -\xi^2(1-\xi) \end{aligned}$$

and two vectors $\mathbf{a}_t$ and $\mathbf{b}_u$

$$\mathbf{a}_t = [\varphi_1(t) \quad \varphi_2(t) \quad h_i\varphi_3(t) \quad h_i\varphi_4(t)],$$

$$\mathbf{b}_u = [\varphi_1(u) \quad \varphi_2(u) \quad l_j\varphi_3(u) \quad l_j\varphi_4(u)].$$

Then the desired value of spline $S(x, y)$ may be calculated by the formula

$$S(x, y) = \mathbf{a}_t \mathbf{Z} \mathbf{b}_u^T,$$

where matrix $\mathbf{Z}$ has the following form:

$$\mathbf{Z} = \begin{pmatrix} z_{ij} & z_{i+1,j} & \mu_{ij}^{(x)} & \mu_{i+1,j}^{(x)} \\ z_{i,j+1} & z_{i+1,j+1} & \mu_{i,j+1}^{(x)} & \mu_{i+1,j+1}^{(x)} \\ \mu_{ij}^{(y)} & \mu_{i+1,j}^{(y)} & \mu_{ij}^{(xy)} & \mu_{i+1,j}^{(xy)} \\ \mu_{i,j+1}^{(y)} & \mu_{i+1,j+1}^{(y)} & \mu_{i,j+1}^{(xy)} & \mu_{i+1,j+1}^{(xy)} \end{pmatrix}$$

2.1.5 Properties of the interpolating bicubic spline functions

A. Approximate property

The problem of approximation of a function defined by the formula

$$z = f(x, y), \quad (x, y) \in R$$

with the help of a function with sufficiently good properties is widely met in applications. For example, such a problem arises if the calculation of the given function's values at the points of the rectangle R involves substantial difficulties.

Problem of the function interpolation. Construct (on the rectangle R) a bicubic spline $S(x, y)$ which coincides with the given function $f(x, y)$ at the knots of the grid ω

$$S(x_i, y_j) = f(x_i, y_j), \quad i = 0, 1, \ldots, m, \quad j = 0, 1, \ldots, n$$

The approximate properties of the interpolating bicubic spline $S(x, y)$ depend on smoothness of the interpolated function $f(x, y)$.

For example, if function $f(x, y)$ is of class $C^4(R)$, then

$$\mid f(x, y) - S(x, y) \mid = \mathcal{O}(h^{3/2} + l^{3/2})$$

where

$$h = \max_{i=0,\ldots,m} h_i, \quad l = \max_{j=0,\ldots,n} l_j$$

(Here $h_i = x_{i+1} - x_i$ and $l_j = y_{j+1} - y_j$ are the widths of grid ω with respect to x and y respectively.)

If function $f(x, y)$ is of class $C^8(R)$, then the approximation order is far above

$$\mid f(x, y) - S(x, y) \mid = \mathcal{O}(h^3 + l^3).$$

Remark. The formulas cited above estimate the order of the approximation error of the given function by the bicubic spline as $h \to 0$ and $l \to 0$ and hold for arbitrary relations between h and l. If the ratio h/l is bounded as $h \to 0$ and $l \to 0$, then the order of the approximation increases by 1,

$$\mid f(x, y) - S(x, y) \mid = \mathcal{O}(h^4 + l^4).$$

B. Extreme property

By analogy with the one-dimensional case we can consider the following variational problem.

Let $R = [a, b] \times [c, d]$ be a given rectangle.

Problem. Construct a function $f(x,y)$ of class $C^{2,2}(R)$ minimizing a functional

$$J(f) = \iint_R \left(\frac{\partial^4 f}{\partial x^2 \partial y^2}\right)^2 dxdy$$

and taking the given values

$$z_{ij}, \qquad i = 0,1,\ldots,m, \quad j = 0,1,\ldots,n$$

at the knots of grid ω

$$a = x_0 < x_1 < \ldots < x_{m-1} < x_m = b,$$

$$c = y_0 < y_1 < \ldots < y_{n-1} < y_n = d.$$

In order that the solution of the problem be unique, additional requirements are necessary. Let us describe some of them.

1. If all the admissible functions satisfy the conditions

$$\begin{array}{lll} \frac{\partial^2 f}{\partial x^2}(x_i,y_j) = 0, & i = 0,m, & j = 0,1,\ldots,n, \\ \frac{\partial^2 f}{\partial y^2}(x_i,y_j) = 0, & i = 0,1,\ldots,m, & j = 0,n, \\ \frac{\partial^4 f}{\partial x^2 \partial y^2}(x_i,y_j) = 0, & i = 0,m, & j = 0,n \end{array}$$

then the solution of the variational problem is uniquely defined and the solution is the interpolating bicubic spline.

2. If the admissible functions satisfy the boundary conditions

$$\begin{array}{lll} \frac{\partial^2 f}{\partial x^2}(x_i,y_j) = z_{ij}^{(1,0)}, & i = 0,m, & j = 0,1,\ldots,n, \\ \frac{\partial^2 f}{\partial y^2}(x_i,y_j) = z_{ij}^{(0,1)}, & i = 0,1,\ldots,m, & j = 0,n, \\ \frac{\partial^4 f}{\partial x^2 \partial y^2}(x_i,y_j) = z_{ij}^{(1,1)}, & i = 0,m, & j = 0,n, \end{array}$$

then the solution of the variational problem is uniquely defined and the solution is the interpolating bicubic spline.

3. If the admissible functions are doubly periodic with periods $T_x = b-a$ and $T_y = d-c$, then the solution of the variational problem is uniquely defined and the solution is the doubly periodic interpolating bicubic spline.

Remark. The splines listed above are the solutions of the formulated problem in a far more wide class of functions, namely, in space $W_2^{2,2}(R)$.

2.1.6 Construction of the interpolating spline surfaces with the help of the interpolating spline functions

The points of the arrays under consideration above are enumerated so that their abscissae and ordinates generate two strongly increased sequences. The fact determines the selection of the approximating class (graphs of functions) and the mode of their construction.

Nonetheless, the method outlined above permits the construction of the interpolating surface with success in a more general case when, as a rule, the numbering of the array points

$$\mathbf{P} = \{\mathbf{P}_{ij}(x_{ij}, y_{ij}, z_{ij}), i = 0, 1, \ldots, m; j = 0, 1, \ldots, n\}$$

and their spatial arrangement are divorced from each other.

It is clear that for the general problem to be solved it is necessary to extend essentially the set of the admissible surfaces by addition of the self-intersecting and self-overlying surfaces, and the surfaces which cannot be projected on any of the coordinate planes. The surfaces from the new set are conveniently described by the parametric equations

$$x = x(u, v), \quad y = y(u, v), \quad z = z(u, v),$$

$$(u, v) \in R = [\alpha, \beta] \times [\gamma, \delta].$$

In addition, let us require that functions $x(u, v)$, $y(u, v)$ and $z(u, v)$ be sufficiently smooth, for example, of class $C^1(R)$ or of class $C^2(R)$.

To find the parametric equations of the surface sequentially passing through all the points of the array we proceed as follows.

1st step. Let us introduce an auxiliary grid

$$\alpha = u_0 < u_1 < \ldots < x_{u-1} < u_m = \beta, \gamma = v_0 < v_1 < \ldots < v_{n-1} < v_n = \delta$$

on arbitrary rectangle

$$R = \{(u, v) | \alpha \le u \le \beta, \quad \gamma \le v \le \delta\}$$

so that the knot's number of the grid coincide with the number of points in array $\mathbf{P}$.

2nd step. For given array $\mathbf{P}$, construct three new auxiliary arrays

$$\begin{aligned}
\mathrm{X} &= \{(u_i, v_j, x_{ij}), i = 0, 1, \ldots, m; j = 0, 1, \ldots, n\}, \\
\mathrm{Y} &= \{(u_i, v_j, y_{ij}), i = 0, 1, \ldots, m; j = 0, 1, \ldots, n\}, \\
\mathrm{Z} &= \{(u_i, v_j, z_{ij}), i = 0, 1, \ldots, m; j = 0, 1, \ldots, n\}.
\end{aligned}$$

3rd step. For each of the arrays X, Y, and Z, find the corresponding interpolating spline functions $x(u, v)$, $y(u, v)$ and $z(u, v)$ by the method described above.

As a consequence we have the parametric equations of the surface passing through all the points $\mathbf{P}_{ij}$.

Remark. The obtained surface is smooth but does not need to be regular because the inequality

$$\operatorname{rank} \begin{pmatrix} x_u & y_u & z_u \\ x_v & y_v & z_v \end{pmatrix} < 2$$

for some $(u^*, v^*) \in (\alpha, \beta) \times (\gamma, \delta)$ is quite possible. In addition, the surface may obtain the self-intersecting lines and self-overlying regions.

2.1.7 Implementation

The above algorithm for construction of the bicubic interpolating spline function given in tabulated form is realized in the program `Splint2` (in Fortran). The routine call looks like

```
        call subroutine Splint2
     *  (m,x,n,y,z,zl,zr,zu,zd,zx,zy,zxy,ind,ibx,
     *  iby,xx,yy,a,b,c,d,u,v, w,t,s,zz,su,sd,spl)
```

Input:
`m` — number of the interpolation knots (along x-axis)
`x` — array (of size `m`) of the interpolation knots (along x-axis)
`n` — number of the interpolation knots (along y-axis)
`y` — array (of size `n`) of the interpolation knots (along y-axis)
`z` — one-dimensional array (of size `m*n`) containing values of the interpolated function at the interpolation knots; the value of the function at knot `(i,j)` is contained in element `z(i+(j-1)*m)`
`ibx` — type of boundary conditions on lines `x=x1` and `x=xm`
`ibx = i (i=1-4)` — conditions of the `i`th type
`zl` — array (of size `n`), element `zl(j)` contains value of the `i`th partial derivative with respect to `x` at knot `(1,j)` for boundary conditions of the `i`th type `(i =1,2)`
`zr` — array (of size `n`), element `zr(j)` contains value of the `i`th partial derivative with respect to `x` at knot `(m,j)` for boundary conditions of the `i`th type `(i =1,2)`
`iby` — type of boundary conditions on lines `y=y1` and `y=yn`
`iby = i (i=1-4)` — conditions of the `i`th type
`zu` — array (of size `m`), element `zu(j)` contains value of the `i`th partial derivative with respect to `y` at knot `(j,n)` for boundary conditions of the `i`th type `(i =1,2)`
`zd` — array (of size `m`) element `zd(j)` contains value of the `i`th partial derivative with respect to `y` at knot `(j,1)` for boundary conditions of the `i`th type `(i =1,2)`
`zxy` — array (of size 4) containing values of mixed derivatives at the corner knots for the boundary conditions of the first or second type
Auxiliary variables:
`a,b,c,d` — work arrays (of size `max(m,n)`)
`s,t,u,v,w,zz,su,sd` — work arrays (of size `max(m,n)+1`)
`zx,zy,zxy` — one-dimensional arrays (of size `m*n`) containing values of the spline parameters
`ind` — mode type:
`ind=0` — the program calculates the spline parameters
`ind=1` — the spline parameters are known
`xx, yy` — coordinate of point where the spline value is calculated

Output:
`spl` — the spline value at point `(xx, yy)`
For illustration of the `Splint2` procedure call, consider the following example.

Example. First construct a grid with 11 × 6 knots in rectangle $[0,\pi] \times [\pi/2]$

$$x_i = \frac{\pi}{10}, \quad i = 0,1,\ldots,10, \quad y_j = \frac{\pi}{10}, \quad j = 0,1,\ldots,5$$

and calculate the values of function $f(x,y) = \sin x \sin y$ at the knots. With these initial data construct the interpolating bicubic spline using the first type boundary conditions and then calculate the values of the given function at points

$$x_i = \frac{\pi}{10}, \quad i = 1,\ldots,21, \quad y_j = \frac{\pi}{10}, \quad j = 1,\ldots,11$$

Denote the set as Ω.

The program for the problem solution may be as follows:

```
      dimension x(100) , y(100) ,  z(100) , zl(100),
     *          zr(100), zu(100),  zd(100), zx(100),
     *          zy(100), zxy(100), u(100) , v(100) ,
     *          a(100) , b(100),   c(100) , d(100) ,
     *          w(100) , t(100),   su(100), sd(100),
     *          s(100) , zz(100)
c
      data pi/3.1415926/
c
      hx  = pi/10.
      hy  = pi/10
      m   = 11
      n   = 6
      ibx = 1
      iby = 1
c
      do i = 1,m
       x(i)  = hx*(i-1)
       zd(i) = sin(x(i))
       zu(i) = 0.
      enddo
c
      do j = 1,n
       y(j)  =  hy*(j-1)
       zl(j) =  sin(y(j))
       zr(j) = -sin(y(j))
      enddo
```

```
c
        do i = 1,m
        do j = 1,n
         z(i+(j-1)*m) = sin(x(i))*sin(y(j))
        enddo
        enddo
c
        zxy(1) =  1.
        zxy(2) = -1.
        zxy(3) =  0.
        zxy(4) =  0.
c
        call  Splint2 (m,x,n,y,z,zl,zr,zu,zd,zx,zy,zxy,0,
     *      ibx,iby,xx,yy,a,b,c,d,u,v,w,t,s,zz,su,sd,spl)
c
        hx = pi/20.
        hy = pi/20.
c
        del = 0.
        do j = 1,11
        do i = 1,21
        xx = hx*(i-1)
        yy = hy*(j-1)
        fun = sin(xx)*sin(yy)
        call  Splint2 (m,x,n,y,z,zl,zr,zu,zd,zx,zy,zxy,
     *     1,ibx,iby,xx,yy,a,b,c,d,u,v,w,t,s,zz,su,sd,spl)
           if (abs(fun-spl).gt.del) del=abs(fun-spl)
        enddo
        enddo
        write (*,*) 'del= ',del
       stop
       end
```

The calculations, with the help of the program, furnish the following result

$$\max |f(x,y) - S(x,y)| = 5.078316E - 05,$$

where maximum is computed on the set Ω.

The source of program `Splint2` may be found on the applied diskette in `splint2.for` file of subdirectory `INTERPOL` where the files containing the other examples using the program are also located. The detailed information on the file names and their contents is presented in **`readme.txt`** file of directory `SPLINES` and in Appendix B.

2.2 Smoothing bicubic splines

2.2.1 Statement of the smoothing problem

Let a grid ω

$$a = x_0 < x_1 < \ldots < x_{m-1} < x_m = b,$$

$$c = y_0 < y_1 < \ldots < y_{n-1} < y_n = d$$

be given in rectangle $R = [a, b] \times [c, d]$. Consider a set of numbers

$$z_{ij}, \qquad i = 0, 1, \ldots, m, \quad j = 0, 1, \ldots, n.$$

Comment on the initial data. The values z_{ij} are convenient to interpret, for example, as the results of measurements of some function $z(x, y)$ for the given values of variables x and y containing a random error. When solving the reconstruction problem of a function by these "experimental" data we see little reason for using the interpolation because the interpolating function will obediently reproduce any oscillations caused by the random component in array $\{z_{ij}\}$. The approach based on the smoothing procedure capable of decreasing the element of randomness in the measurement results is more natural.

There exist many various ways of smoothing. More often the methods are formulated as the extremum problems for the given functional defined by grid ω and array $\{z_{ij}\}$ on some class of the admissible functions.

2.2.2 Definition of the smoothing bicubic spline function

The function $S(x, y)$ defined on grid ω is called *a smoothing bicubic spline function* if the function

1) is a bicubic polynomial

$$S(x, y) = \sum_{p=0}^{3} \sum_{q=0}^{3} a_{p,q}^{(i,j)} (x - x_i)^p (y - y_j)^q$$

in each cell

$$R_{ij} = \{(x, y) \mid x_i \le x \le x_{i+1}, \quad y_j \le y \le y_{j+i}\},$$

$$i = 0, 1, \ldots, m, \quad j = 0, 1, \ldots, n$$

2) is a function of class $C^{2,2}(R)$,

3) minimizes the functional

$$J(f) = \int_a^b \int_c^d \left(\frac{\partial^4 f(x, y)}{\partial x^2 \partial y^2} \right)^2 dx dy +$$

$$+\sum_{i=0}^{m}\frac{1}{\rho_i}\int_a^b\left(\frac{\partial^2 f(x_i,y)}{\partial y^2}\right)^2 dy+\sum_{j=0}^{n}\frac{1}{\sigma_j}\int_c^d\left(\frac{\partial^2 f(x,y_j)}{\partial x^2}\right)^2 dx$$

$$+\sum_{i=0}^{m}\sum_{j=0}^{n}\frac{1}{\rho_i\sigma_j}\left(f(x_i,y_j)-z_{ij}\right)^2,$$

where z_{ij} $(i=0,1,\ldots,m; j=0,1,\ldots,n)$, $\rho_i>0$ $(i=0,\ldots,m)$ and $\sigma_j>0$ $(j=0,\ldots,n)$ are the given numbers, and

4) satisfies the boundary conditions of one of three types described below.

2.2.3 Boundary conditions

The boundary conditions are given in the form of the restrictions on the values of spline and its derivatives at the boundary and corner knots of grid ω.

Boundary conditions of the first type

The values of the first partial derivatives of the desired function with respect to x and y are given at the boundary knots of grid ω and the values of the mixed second derivative at the corner knots

$$\begin{array}{lll}\frac{\partial S}{\partial x}(x_i,y_j)=z_{ij}^{(x)}, & i=0,m, & j=0,1,\ldots,n,\\ \frac{\partial S}{\partial y}(x_i,y_j)=z_{ij}^{(y)}, & i=0,1,\ldots,m, & j=0,n,\\ \frac{\partial^2 S}{\partial x\partial y}(x_i,y_j)=z_{ij}^{(xy)}, & i=0,m, & j=0,n.\end{array}$$

The total number of first type boundary conditions is equal to $2m+2n+8$.

A way of the spline definition by the boundary conditions of the first type is illustrated in Figure 4. The knots where the values z_{ij} of spline S are given are marked off by the bullets $\bullet$, the knots where the values of the first partial derivatives S_x and S_y are given are marked off by horizontal $\triangleleft$, $\triangleright$ and vertical $\triangle$, $\bigtriangledown$ arrows, respectively, and the knots where the values of the mixed derivative S_{xy} are given are marked off by rhombus $\diamond$.

Boundary conditions of the second type

The values of the second partial derivatives of the desired function with respect to x and y are given at the boundary knots of grid ω and the values of the mixed derivative S_{xxyy} at the corner knots:

$$\begin{array}{lll}\frac{\partial^2 S}{\partial x^2}(x_i,y_j)=0, & i=0,m, & j=0,1,\ldots,n,\\ \frac{\partial^2 S}{\partial y^2}(x_i,y_j)=0, & i=0,1,\ldots,m, & j=0,n,\\ \frac{\partial^4 S}{\partial x^2\partial y^2}(x_i,y_j)=0, & i=0,m, & j=0,n.\end{array}$$

The total number of the second type boundary conditions is equal to $2m + 2n + 8$.

A way of the spline definition by the boundary conditions of the first type is illustrated in Figure 4. The knots where the values z_{ij} of spline S are given are marked off by the bullets $\bullet$, the knots where the values of the second partial derivatives S_{xx} and S_{yy} are given are marked off by horizontal $\triangleleft$, $\triangleright$ and vertical $\triangle$, $\triangledown$ arrows, respectively, and the knots where the values of the mixed derivative S_{xxyy} are given are marked off by rhombus $\diamond$.

Boundary conditions of the third type

are called *periodic*. In this case the spline $S(x, y)$ and its partial derivatives must be doubly periodic; their period with respect to variable x is equal to $T_x = b - a$ and with respect to variable y is equal to $T_y = d - c$

$$\begin{array}{lll}
S(x_0, y_j) = S(x_m, y_j), & j = 0, 1, \ldots, n & \\
S(x_i, y_0) = S(x_i, y_n), & i = 0, 1, \ldots, m & \\
\frac{\partial^k S}{\partial x^k}(x_0, y_j) = \frac{\partial^k S}{\partial x^k}(x_m, y_j), & j = 0, 1, \ldots, n, & k = 1, 2, \\
\frac{\partial^l S}{\partial y^l}(x_i, y_0) = \frac{\partial^l S}{\partial y^l}(x_i, y_n), & i = 0, 1, \ldots, m, & l = 1, 2, \\
\frac{\partial^{2k} S}{\partial x^k \partial y^k}(x_0, y_j) = \frac{\partial^{2k} S}{\partial x^k \partial y^k}(x_m, y_j), & j = 0, 1, \ldots, n, & k = 1, 2, \\
\frac{\partial^{2k} S}{\partial x^k \partial y^k}(x_i, y_0) = \frac{\partial^{2k} S}{\partial x^k \partial y^k}(x_i, y_n), & i = 0, 1, \ldots, m, & k = 1, 2.
\end{array}$$

Theorem. Among all the functions of class $C^{2,2}(R)$ satisfying the boundary conditions of the second type (*natural conditions*) it is precisely a bicubic spline that minimizes functional $J(f)$.

Definition. A bicubic spline minimizing functional $J(f)$ and satisfying the boundary conditions of the ith type is called *a smoothing bicubic spline of the ith type.*

Remark. In addition to the described conditions there exist so-called mixed boundary conditions, that is, conditions of the distinct types for the distinct variables. In this case it is necessary to correlate these conditions: if the first type conditions are given with respect to variable x, and the second type conditions with respect to variable y, then at the vertices of the rectangle R we need to take partial derivatives S_{xyy}, etc.

2.2.4 Construction of the smoothing bicubic spline function

The algorithm of the construction of the smoothing bicubic spline is based on the fact that the construction may be reduced to the sequence of the one-dimensional smoothing problems.

1st step of the algorithm

On every line $y = y_k$ $(k = 0, 1, \ldots, n)$ we construct the one-dimensional smoothing spline $S_k(x)$ with respect to variable x using the corresponding end conditions (the procedure of the one-dimensional spline construction is described in the first chapter).

2nd step of the algorithm
For every knot (x_i, y_j) of grid ω we calculate "corrected" values $z_{ij}^{(1)}$ by formula $z_{ij}^{(1)} = S_j(x_i)$.

3rd step of the algorithm
The third step is similar to the 1st step: on each of lines $x = x_l$, $(l = 0, 1, \ldots, m)$ we construct the smoothing cubic (with respect to variable y) spline $\tilde{S}_l(y)$ generated by array $z_{lj}^{(1)}$ and satisfying the boundary conditions of the corresponding type.

4th step of the algorithm
For every knot of grid ω we calculate new "corrected" values $z_{ij}^{(2)}$ by formula $z_{ij}^{(2)} = \tilde{S}_i(y_j)$.

5th step of the algorithm
For the new set of values $z_{ij}^{(2)}$ we construct the interpolating spline $S(x, y)$ and the spline is a desired smoothing spline.

After the 5th step of the algorithm is over, we know magnitudes

$$z_{ij}^{(2)}, \quad \mu_{ij}^{(x)}, \quad \mu_{ij}^{(y)}, \quad \mu_{ij}^{(xy)}$$

at every knot of grid ω. The numbers define completely the smoothing bicubic spline satisfying the given boundary conditions.

Algorithm of the calculation of the spline values by the given magnitudes

$$z_{ij}^{(2)}, \quad \mu_{ij}^{(x)}, \quad \mu_{ij}^{(y)}, \quad \mu_{ij}^{(xy)}$$

coincides with the algorithm described in the subsection 2.1.4.

Remark 1. The splines enumerated above are the solutions of the corresponding variation problems in a far more wide class of functions, namely, in space $W_2^{2,2}(R)$.

Remark 2. With the help of the weight factors ρ_i and σ_j one can control the behavior of the smoothing spline, for example, defining the layer shape near the "experiment" points z_{ij} where the spline must be. Since the weight factors ρ_i and σ_j appear in the double sum of functional $J(f)$ in the form of product, their choice influences the character of smoothing not at one point (x_i, y_j) but on lines $x = x_i$ and $y = y_j$.

Remark 3. A far more complex case when the weight factors in the double sum have a form $\tau_{ij} \neq \rho_i\sigma_j$ is not dealt with in this book.

2.2.5 Construction of the smoothing spline surfaces with the help of the smoothing spline functions

The points of the arrays under consideration above are enumerated so that their abscissae and ordinates generate two strongly increasing sequences. The fact determines the choice of the approximating class of surfaces (graphs of functions) and the mode of their construction.

Nonetheless, the method outlined above permits one to construct the smoothing surface with much success in a more general case when the numbering of the array points

$$\mathbf{P} = \{\mathbf{P}_{ij}(x_{ij}, y_{ij}, z_{ij}), i = 0, 1, \ldots, m; j = 0, 1, \ldots, n\}$$

and their arrangement are, as a rule, divorced from each other.

It is clear that for the general problem to be solved it is necessary to extend essentially the set of the admissible surfaces by addition of the self-intersecting and self-overlying surfaces, and the surfaces which cannot be projected on any of the coordinate planes. The surfaces from the new set are conveniently described by the parametric equations

$$x = x(u, v), \quad y = y(u, v), \quad z = z(u, v),$$

$$(u, v) \in R = [\alpha, \beta] \times [\gamma, \delta].$$

In addition, let us require that functions $x(u, v)$, $y(u, v)$ and $z(u, v)$ be sufficiently smooth, for example, of class $C^1(R)$ or of class $C^2(R)$.

To find the parametric equations of the surface smoothing out the given array, we proceed as follows.

1st step. Let us introduce an auxiliary grid

$$\alpha = u_0 < u_1 < \ldots < x_{u-1} < u_m = \beta, \gamma = v_0 < v_1 < \ldots < v_{n-1} < v_n = \delta$$

on arbitrary rectangle

$$R = \{\alpha \le u \le \beta, \quad \gamma \le v \le \delta\}$$

so that the knot's number of the grid coincides with the number of points in the array $\mathbf{P}$.

2nd step. For given array $\mathbf{P}$ construct three auxiliary arrays

$$\begin{aligned} \mathrm{X} &= \{(u_i, v_j, x_{ij}), i = 0, 1, \ldots, m; j = 0, 1, \ldots, n\}, \\ \mathrm{Y} &= \{(u_i, v_j, y_{ij}), i = 0, 1, \ldots, m; j = 0, 1, \ldots, n\}, \\ \mathrm{Z} &= \{(u_i, v_j, z_{ij}), i = 0, 1, \ldots, m; j = 0, 1, \ldots, n\}. \end{aligned}$$

3rd step. For each of the arrays X, Y, and Z, find the corresponding smoothing spline functions $x(u, v)$, $y(u, v)$ and $z(u, v)$ by the methods described above.

As a consequence we have the parametric equations of the smoothing spline surface generated by points $\mathbf{P}_{ij}$.

Remark. The obtained surface is smooth but does not need to be regular because the inequality

$$\operatorname{rank}\begin{pmatrix} x_u & y_u & z_u \\ x_v & y_v & z_v \end{pmatrix} < 2$$

for some $(u^*, v^*) \in (\alpha, \beta) \times (\gamma, \delta)$ is quite possible. In addition, the surface may obtain the self-intersecting lines and the self-overlying regions.

2.2.6 Implementation

The above algorithm for the construction of the smoothing bicubic spline function and calculation of its values at the arbitrary point is realized in the program Smspl2 (in Fortran). The routine call looks like

```
      subroutine Smspl2
   *  (m,x,n,y,z,zl,zr,zu,zd,rho,sgm,zx,zy,zxy,ind,ib,xx,
   *  yy,a,b,c,d,e,g,u,v, w,t,s,zz,su,sd,p,q,r,dd,zm,spl)
```

Input:
m — number of knots (along x-axis)
x — array (of size m) containing knots (along x-axis)
n — number of knots (along y-axis)
y — array (of size n) containing knots (along y-axis)
z — one-dimensional array (of size m*n) containing values of the smoothed function at the knots; value of function at knot (i,j) is contained in element z(i+(j-1)*m)
ib — type of boundary conditions on lines x=x1 and x=xm and on lines y=y1 and y=yn
ib = i (i=1-3) — conditions of ith type
zl — array (of size n), element zl(j) contains value of the first partial derivative with respect to x at knot (1,j) for the boundary conditions of the first type
zr — array (of size n), element zr(j) contains value of the first partial derivative with respect to x at knot (m,j) for the boundary conditions of the first type
zu — array (of size m), element zu(j) contains value of the first partial derivative with respect to y at knot (j,m) for the boundary conditions of the first type
zd — array (of size m), element zd(j) contains value of the first partial derivative with respect to y at knot (j,1) for the boundary conditions of the first type
zxy — array (of size 4) containing values of mixed derivatives at the corner knots for the boundary conditions of the first type

`rho` — array (of size `m`) containing values of the weight coefficients
`sgm` — array (of size `n`) containing values of the weight coefficients
`a,b,c,d,su,sd` — work arrays (of size `max(m,n)`)
`s,t,u,u,v` — work arrays (of size `max(m,n)+1`)
`p,q,r,e,g` — work arrays (of size `max(m,n)+2`)
`zx,zy,zm,zz,dd` — work arrays (of size `m*n`)
`ind` — mode type:
`ind=0` — program calculates spline's parameters
`ind=1` — the spline's parameters are known
`xx,yy` — coordinates of point where the spline value is calculated
Output:
`spl` — the spline value at point `(xx,yy)`.
Program `Smspl2` possesses a call to routines `Smspline` and `Splint2`.

For illustration of the `Smspl2` procedure call consider the following example simulating the smoothing of the experiment data.

Suppose that function $z = \sin(x + y^2)$ is "measured" at the knots of grid

$$\omega = \{x_i = \frac{1}{10}i, \quad y_j = \frac{1}{20}j, \quad i = 0, 1, \ldots, 5, \quad j = 0, 1, \ldots, 20\}$$

with error ξ distributed uniformly on segment [0.1,0.1].

Let us construct the smoothing and interpolating splines satisfying the same boundary conditions of the first type and compare their behavior on rectangle $[0, 0.5 \times [0, 1]$.

Take magnitudes

$$\rho_i = 0.001, \quad i = 0, \ldots, 5, \quad \sigma_j = 0.001, \quad j = 0, \ldots, 20$$

as weight coefficients.

The program for the problem solution may be as follows:

```
      dimension x(200),  y(200),    z(200),    zl(200),
     *          zr(200), zu(200),   zd(200),   zx(200),
     *          zy(200), zxy(200),  u(200),    v(200),
     *          a(200),  b(200),    c(200),    d(200),
     *          w(200),  t(200),    su(200),   sd(200),
     *          s(200),  zz(200),   rho(200),  sgm(200),
     *          p(200),  q(200),    r(200),    dd(200),
     *          zm(200), e(200),    g(200)
c
      data rho / 200*0.001/
      data sgm / 200*0.001/
c
      m  = 6
      n  = 21
```

```
      hx = 0.5/(m-1)
      hy = 1./(n-1)
c
      do i = 1,m
         x(i) = hx*(i-1)
         zu(i)= 2.*cos(1. + x(i))
         zd(i)= 0.
      enddo
c
      do j = 1,n
         y(j) = hy*(j-1)
         zl(j)= cos(y(j)**2)
         zr(j)= cos(0.5 + y(j)**2)
      enddo
c
      zxy(1) = 0.
      zxy(2) = 0.
      zxy(3) = -2.*sin(1.)
      zxy(4) = -2.*sin(1.5)
      call seed(1234)
      do j = 1,n
      do i = 1,m
         call random(df)
         z(i+(j-1)*m)=sin(x(i)+y(j)**2)+0.4*(df-0.5)
      enddo
      enddo
c
      ind = 0
      ib  = 1
c
      call  Smspl2  (m,x,n,y,z,zl,zr,zu,zd,rho,sgm,zx,
     *                zy,zxy,ind,ib,xx,yy,a,b,c,d,e,g,
     *                u,v,w,t,s,zz,su,sd,p,q,r,dd,zm,spl)
c
      ind=1
      open (1,file='res1.dat')
      hx = hx/2.
      hy = hy/5.
c
      do i = 1,11
        xx = hx*(i-1)
         do j = 1,101
          yy = hy*(j-1)
          call Smspl2 (m,x,n,y,z,zl,zr,zu,zd,rho,sgm,
     *                   zx,zy,zxy,ind,ib,xx,yy,a,b,c,
```

```
     *                      d,e,g,u,v, w,t,s,zz,su,sd,p,q,
     *                      r,dd,zm,spl)
               write (1,*)  xx, yy,  spl
              enddo
           enddo
c
          stop
          end
```

Remark. A comparison of the interpolating and smoothing splines satisfied the same boundary conditions (of the first type) shows that the oscillation peculiar to the interpolating spline nearly completely disappeared when we use the smoothing spline.

The source of program `Smspl2` may be found on the applied diskette in `smspl2.for` file of subdirectory `SMOOTH` where the files containing the other examples using the program are also located. The detailed information on the file names and their contents is presented in `readme.txt` file of directory `SPLINES` and in Appendix B.

Part II
Geometric splines

In many cases the requirement for the constructed curve or surface to be projected uniquely on a straight line or a plane is too rigid. Extending the admissible classes of curves and surfaces, it is natural to choose a more general way of their fragments' description as well. As a new way for such description it is convenient to take the parametric method.

The parametric description of the curves and surfaces offers definite advantage over the other methods, in particular, it doesn't impose practically any restrictions on the set of the support vertices.

At the same time, this method requires an extreme caution: for a composite curve or surface described by the parametric equations to be sufficiently regular it is necessary to be very careful, especially at the joining sites. The point is that the connection between the geometric and functional properties is not so simple.

A choice of the cubic polynomials for the coordinate functions of the constructed curves is well founded: the coordinate functions must be relatively simple and, at the same time, provide reasonable smoothness.

Here, as in the first part, it is reasonable to use the results of solving the one-dimensional problems in the construction of the corresponding composite surfaces, in particular, to restrict the consideration to the polynomial description of the coordinate functions. This simplifies solving the problem of creating the complex-shaped surfaces.

Chapter 3

Spline curves

The general problem of this chapter may be formulated as follows: construct a smooth curve for a given array of vertices

$$\mathbf{P} = \{\mathbf{P}_0, \mathbf{P}_1, \ldots, \mathbf{P}_{m-1}, \mathbf{P}_m\}$$

accounting for their enumeration so that the curve passes near these vertices and is subject to some additional conditions. The conditions may be of various character. For example, one can require the desired curve to pass through all of the given vertices, or, passing through each of the given vertices to be tangent to given directions, or to be closed, etc.

In the search for the appropriate solution of the problem, a polygon, the segments of which link the neighboring vertices of the given array, plays an important role. The polygon is called *a control polygon* or *a support polygon* and its vertices are called *control* or *support vertices* (Figure 1). Every vertex of the given array is either *an inner vertex* or *an end vertex*. The vertices $\mathbf{P}_i$ $(i = 1, \ldots, m-1)$ of the array $\mathbf{P}$ are inner points and vertices $\mathbf{P}_0$ and $\mathbf{P}_m$ are endpoints. In many cases the control polygon discloses the course of an unknown curve sufficiently well. The property is of prime importance for solving the smoothing problem.

Contrary to the case considered in the first chapter, here, we have no restrictions on the set of vertices — the set may be planar or spartial, the mutual arrangement of the vertices may be absolutely arbitrary, some of the vertices may coincide, etc. Because of this, the description of a desired curve should be searched in a more general parametric form, for example, as follows:

$$\mathbf{R}(t) = \sum a_i(t)\mathbf{P}_i, \qquad (*)$$

where $a_i(t)$ are functional coefficients to be defined.

If the vertex number in the given array $\mathbf{P}$ is large enough, searching for the universal functional coefficients $a_i(t)$, as a rule, may be rather difficult. Even though the universal coefficients $a_i(t)$ are found, often it turns out

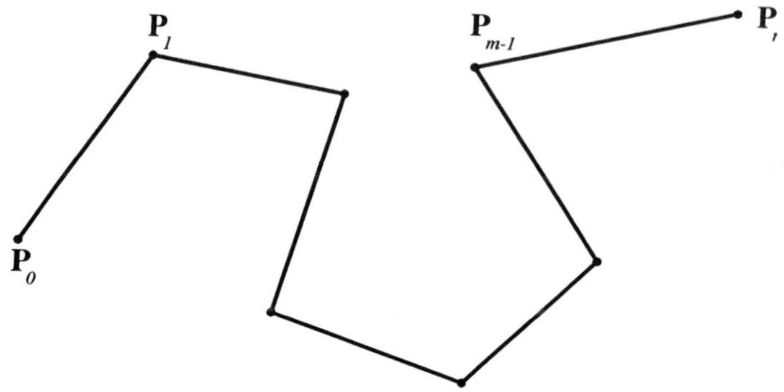

Figure 1.

that, in parallel with the wanted properties, they have the properties which are not easily compatible with the expected behavior of the curve (for example, the curve described by equation (*) with these coefficients may oscillate or deviate greatly from the given set).

To solve the formulated problem, we can conveniently use the curves composed of the elementary fragments. If the elementary fragments are constructed according to a unified relatively simple scheme, the corresponding composite curves are called *spline curves*.

The parametric equations of every elementary fragment are sought in the form (*) under condition that the scalar functional coefficients are of the same nature (for example, polynomials of the same degree). But for constructing a concrete fragment of the desired curve, we use only part of the given vertices from the set **P**.

The polynomial coefficients of the small (second or third) degree are usually used for the description of the elementary curves and their geometric properties (the information is essential for joining). The reason for this is that the coefficients are calculated relatively easily. Of course, with the help of the polynomials of the higher orders one can describe the complex-shaped curves. But such polynomials have many coefficients whose geometric and physical meaning is very difficult to understand. In addition, using the polynomials of the higher degrees may initiate the undesired oscillations of the resulting curve.

The most widespread ways of the composite curve construction are the methods involving the use of cubic polynomials (by the way, the polynomials were extensively used in the first chapter). The choice accounts for both the differential and external-geometric requirements imposed on the desired curve.

The above defines the way the chapter's sections are presented. The chapter begins with the description of the minimal necessary information from the differential geometry of curves. Then we consider various curve classes which help resolve the approximation problem formulated above. The description of each class is performed practically following the same scheme. First we define elementary curves (fragments), and then describe how the elementary curves must be joined in order to obtain a composite curve with sufficiently good characteristics. Much attention is devoted to the description of the main properties of the curves under consideration. The approach will help a user in the choice of the most appropriate class of curves. Part of the properties is formulated in the form of the answers to a standard set of questions, including such questions as:

1) What is the smoothness order of the constructed curve?
2) What are the distinguishing features of the curve arrangement with respect to the given vertices of the control polygon?
3) How does the change of the support vertices influence the curve form?

It is useful to know how the curve shape and the simplest geometric transformations (affine or perspective) are related to each other. Let us dwell on these connections more closely.

It is clear that the constructed curve may be subjected to any (affine or perspective) transformation. Another approach is to first transform the given array of the support vertices and only then use the data received to construct a curve of the considered class.

If the results coincide, the class of curves is said to be *invariant with respect to the (affine or perspective) transformation.*

Recall that the affine transformations include a rotation, a dilation, a translation and their combinations, and the projective transformations as well as a transformation of the perspective.

Most of the classes under consideration possess the affine invariance property. Among them there exist the projective invariant ones. While the final result is independent of the fulfillment sequence of these two operations (transformation of the support vertices and construction of the curve), nevertheless the choice of the operation sequence may substantially reduce the volume of calculations. It is easy to see, for example, that a rotation of the control array around an axis and a subsequent construction of the curve require a smaller volume of calculation than the rotation of the constructed curve.

Every so often the necessity arises to represent the constructed three-dimensional object (in this case, the curve) on the display. For the projective invariant curve, it is more useful first to project the control array on the plane and after that construct the curve.

Use of the rational curves introduces the wanted adaptability into solution of the approximation problem because the free numerical parameters take part in description of the curves. The possibility of choice permits one to take into account many circumstances which inevitably arise in the curve construction with the properties prescribed. At the same time, the rational curves hold many of the properties of the corresponding nonrational (polynomial) curves. The projective invariance of the rational curves is their most attractive property (very useful in visualization).

For a composite curve, the question of the (whole) parametrization is of importance. The problem is that the connection between the analytic description of the composite curve and its geometric shape is not necessarily simple and easy, and some of problems which can arise under the curve composition are illustrated by some typical examples.

As a rule, each of the sections devoted to a definite class of the spline curves is completed with a program implementation and specific calculating examples.

3.1 Main facts of the curve theory

3.1.1 Parametrized curves

A set γ of points $M(x, y, z)$ whose Cartesian coordinates x, y, and z are defined by formulas

$$x = x(t), \quad y = y(t), \quad z = z(t), \quad a \le t \le b,$$

where functions $x(t)$, $y(t)$, and $z(t)$ are continuous on segment $[a, b]$, or in *vector-matrix* form

$$\mathbf{R} = \mathbf{R}(t) = \begin{pmatrix} x(t) \\ y(t) \\ z(t) \end{pmatrix}, \quad a \le t \le b$$

is called *a parametrically defined (parametrized) space curve.*

The relations are called *parametric equations of space curve* γ, or simply *a parametrization of curve* γ.

3.1.2 Smooth and regular curves

We say that the space curve γ is *a C^r-smooth curve with respect to the given parametrization* if vector function $\mathbf{R}(t)$ is C^r-smooth on segment $[a, b]$. That is, each of the coordinate functions $x(t)$, $y(t)$, and $z(t)$ has continuous derivatives up to and including r on the segment (we consider right-side and left-side derivatives at the end points a and b, respectively).

A smooth parametrization is called *regular* if

$$\dot{\mathbf{R}}(t) = \begin{pmatrix} \dot{x}(t) \\ \dot{y}(t) \\ \dot{z}(t) \end{pmatrix} \ne \mathbf{O}, \quad a \le t \le b,$$

where the differentiation with respect to parameter t is denoted by a point.

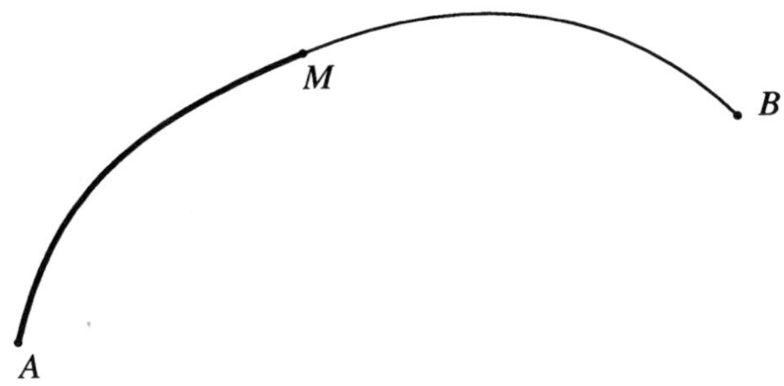

Figure 2.

A value

$$| \dot{\mathbf{R}}(t) | = \sqrt{\dot{x}^2(t) + \dot{y}^2(t) + \dot{z}^2(t)}$$

is called *a (scalar) velocity of the curve* (with respect to the given parametrization) at point $M = M(t)$, and

$$s = s(t) = \int_a^t | \dot{\mathbf{R}}(\xi) | d\xi = \int_a^t \sqrt{\dot{x}^2(\xi) + \dot{y}^2(\xi) + \dot{z}^2(\xi)} d\xi$$

is called *an arc length* of the curve segment $\smile AM$ (Figure 2).

3.1.3 Transformation of parameter

A transformation of parameter

$$\tau = \tau(t),$$

where

$$\dot{\tau}(t) > 0, \quad a \le t \le b$$

changes the parametric equations of curve γ

$$\tilde{\mathbf{R}}(\tau) = \mathbf{R}(t(\tau)), \quad \alpha \le \tau \le \beta$$

(here, function $t = t(\tau)$ is the inverse of function $\tau = \tau(t)$) and its velocity

$$\left| \frac{d\tilde{\mathbf{R}}}{d\tau} \right| = | \dot{\mathbf{R}} | \left| \frac{dt}{d\tau} \right|$$

but holds the curve shape and the curve smoothness (the latter is held when $\tau(t)$ is a sufficiently smooth function).

A curve parametrization with an arc length as a parameter is called *a natural parametrization* and the arc length *a natural parameter.* A scalar velocity of the naturally parametrized curve is equal to 1 everywhere

$$| \mathbf{R}'(s) | = \sqrt{[x'(s)]^2 + [y'(s)]^2 + [z'(s)]^2} = 1$$

(Here, prime $'$ denotes the differentiation with respect to the arc length.)

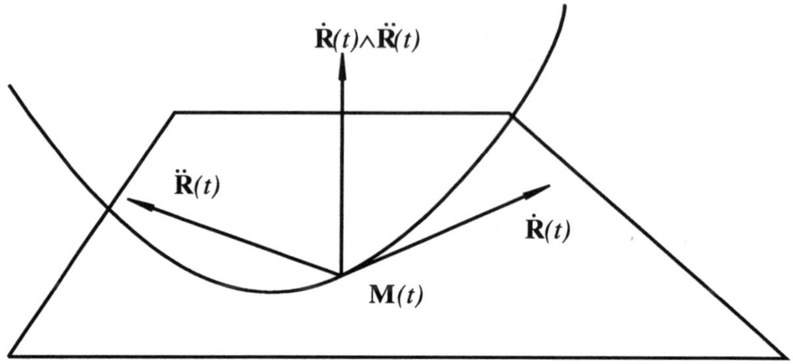

Figure 3.

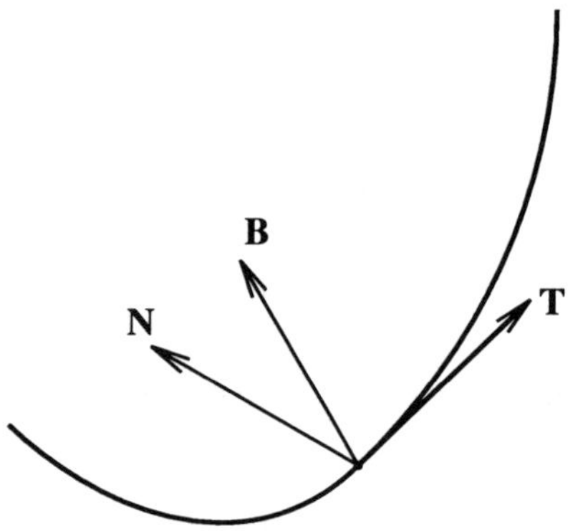

Figure 4.

3.1.4 The Frénet trihedral

Let $\mathbf{R} = \mathbf{R}(t)$ be a parametric vector equation of C^3-regular curve γ.

Then

$$\mathbf{T}(t) = \frac{\dot{\mathbf{R}}(t)}{\mid \dot{\mathbf{R}}(t) \mid}$$

is *a unit tangent vector* of the curve at point $M(t)$.

Let vectors $\dot{\mathbf{R}}(t)$ and $\ddot{\mathbf{R}}(t)$ be noncollinear.

Plane Π passing through point $M(t)$ parallel to vectors $\dot{\mathbf{R}}(t)$ and $\ddot{\mathbf{R}}(t)$ is called *an osculating plane* of curve γ at the point, and its normal vector $\dot{\mathbf{R}}(t) \wedge \ddot{\mathbf{R}}(t)$ is a direction vector of *a binormal* (Figure 3).

The following two vectors play an important role in the curve theory: *a unit binormal vector*

$$\mathbf{B}(t) = \frac{\dot{\mathbf{R}}(t) \wedge \ddot{\mathbf{R}}(t)}{\mid \dot{\mathbf{R}}(t) \wedge \ddot{\mathbf{R}}(t) \mid}$$

and *a unit principal normal vector*

$$\mathbf{N}(t) = \mathbf{B}(t) \wedge \mathbf{T}(t).$$

(It lies in the osculating plane.)

A triple of the mutual perpendicular unit vectors

$$\mathbf{T}(t), \quad \mathbf{N}(t), \quad \mathbf{B}(t)$$

is called *a basic trihedral* or *a trihedral of Frénet* (Figure 4).

If a curve is defined by the parametric vector equation $\mathbf{R} = \mathbf{R}(t)$, the vectors of the Frénet trihedral may be calculated by formulas

$$\mathbf{T}(t) = \frac{\dot{\mathbf{R}}(t)}{\mid \dot{\mathbf{R}}(t) \mid}, \quad \mathbf{N}(t) = \frac{[\dot{\mathbf{R}}(t) \wedge \ddot{\mathbf{R}}(t)] \wedge \dot{\mathbf{R}}(t)}{\mid [\dot{\mathbf{R}}(t) \wedge \ddot{\mathbf{R}}(t)] \wedge \dot{\mathbf{R}}(t) \mid},$$

$$\mathbf{B}(t) = \frac{\dot{\mathbf{R}}(t) \wedge \ddot{\mathbf{R}}(t)}{\mid \dot{\mathbf{R}}(t) \wedge \ddot{\mathbf{R}}(t) \mid}.$$

In case of the natural parametrization $\mathbf{R} = \mathbf{R}(s)$, the formulas for these vectors may be written as follows:

$$\mathbf{T}(s) = \mathbf{R}'(s), \quad \mathbf{N}(s) = \frac{\mathbf{R}''(s)}{\mid \mathbf{R}''(s) \mid}, \quad \mathbf{B}(s) = \frac{\mathbf{R}'(s) \wedge \mathbf{R}''(s)}{\mid \mathbf{R}'(s) \wedge \mathbf{R}''(s) \mid}.$$

The Frénet trihedral is independent of the curve parametrization.

Remark. As a rule, the formulas of the curve theory involving the natural parameter appear more simple and disclose their own geometric significance more explicitly, but in applications the use of the natural parametrization for the curve construction is made difficult by the absence of the curve itself.

3.1.5 Curvature and torsion of the curve

For any C^3-regular curve the unit vectors of tangent $\mathbf{T}(s)$, principal normal $\mathbf{N}(s)$ and binormal $\mathbf{B}\ (s)$ are connected at every point by *the Serret-Frénet* formula

$$\begin{pmatrix} \mathbf{T}'(s) \\ \mathbf{N}'(s) \\ \mathbf{B}'(s) \end{pmatrix} = \begin{pmatrix} 0 & k_1(s) & 0 \\ -k_1(s) & 0 & k_2(s) \\ 0 & -k_2(s) & 0 \end{pmatrix} \begin{pmatrix} \mathbf{T}(s) \\ \mathbf{N}(s) \\ \mathbf{B}(s) \end{pmatrix}$$

where $k_1(s)$ is *a curvature* and $k_2(s)$ is *a torsion* of the curve at the point $M(s)$.

The curvature and torsion of an arbitrary regular curve may be calculated by the following formulas:

$$k_1 = k_1(s) = \mid \mathbf{R}''(s) \mid, \quad k_1 = k_1(t) = \frac{\mid \dot{\mathbf{R}}(t) \wedge \ddot{\mathbf{R}}(t) \mid}{\mid \dot{\mathbf{R}}(t) \mid^3}$$

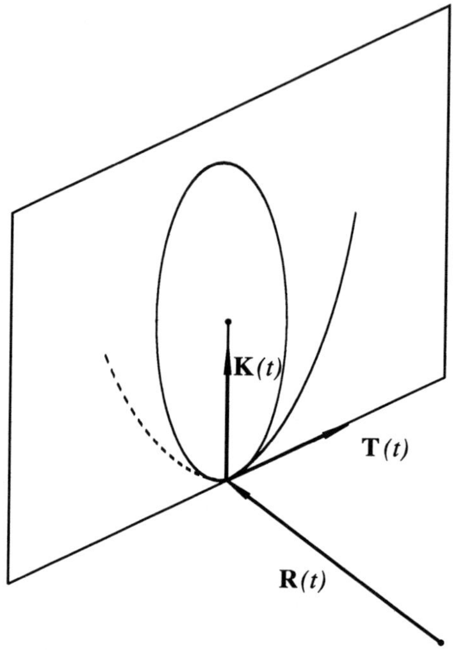

Figure 5.

and

$$k_2 = k_2(s) = \frac{(\mathbf{R}'(s) \wedge \mathbf{R}''(s) \wedge \mathbf{R}'''(s))}{\mid \mathbf{R}''(s) \mid^3},$$

$$k_2 = k_2(t) = \frac{(\dot{\mathbf{R}}(t) \wedge \ddot{\mathbf{R}}(t) \wedge \dddot{\mathbf{R}}(t))}{(\dot{\mathbf{R}}(t) \wedge \ddot{\mathbf{R}}(t))^3}.$$

The curvature k_1 and the torsion k_2 are independent of the choice of the curve parametrization and disclose the warping degree of the curve: the curvature describes the degree of the curve deflection from a straight line (the curvature of a straight line is identically equal to zero) and the torsion — the degree of the curve deflection from a plane (the torsion of a planar regular curve is identically equal to zero).

A vector calculated by formulas

$$\mathbf{K} = \mathbf{K}(s) = \mathbf{R}''(s), \quad \mathbf{K} = \mathbf{K}(t) = \frac{[\dot{\mathbf{R}}(t) \wedge \ddot{\mathbf{R}}(t)] \wedge \dot{\mathbf{R}}(t)}{\mid \dot{\mathbf{R}}(t) \mid^4}$$

is called *a curvature vector* of curve γ. The vector lies in the osculating plane of the curve at point $M(t)$, is perpendicular to the curve tangent at the point, and its length is equal to the curvature of the curve at point $M(t)$, $\mid \mathbf{K} \mid = k_1$.

A circle with a center at point

$$\mathbf{R}(t) + \frac{1}{k_1(t)}\mathbf{N}(t) = \mathbf{R}(t) + \frac{1}{k_1^2(t)}\mathbf{K}(t)$$

and radius $1/k_1(t)$ is called *an osculating circle* of curve γ at point $M(t)$. The osculating circle lies on the osculating plane of curve γ at point $M(t)$ and *an order of osculation (an order of contact)* of the circle and the curve is not below two. The center of the osculating circle is called *a center of curvature* and its radius *a radius of curvature* of the curve at point $M(t)$ (Figure 5).

3.1.6 Planar curves

A. Parametric description

Let us write the principal formulas for the case when the curve γ is planar,

$$\mathbf{R} = \mathbf{R}(t) = \begin{pmatrix} x(t) \\ y(t) \end{pmatrix}, \quad a \le t \le b$$

a curve velocity at point $M(t)$,

$$|\,\dot{\mathbf{R}}(t)\,| = \sqrt{\dot{x}^2(t) + \dot{y}^2(t)}$$

a unit tangent vector,

$$\mathbf{T}(s) = \mathbf{R}'(s), \quad \mathbf{T}(t) = \frac{1}{|\,\dot{\mathbf{R}}(t)\,|}\begin{pmatrix} \dot{x}(t) \\ \dot{y}(t) \end{pmatrix}$$

a unit normal vector (Figure 6),

$$\mathbf{N}(s) = \frac{\mathbf{R}''(s)}{|\,\mathbf{R}''(s)\,|}, \quad \mathbf{N}(t) = \frac{1}{|\,\dot{\mathbf{R}}(t)\,|}\begin{pmatrix} -\dot{y}(t) \\ \dot{x}(t) \end{pmatrix}$$

a curvature of the curve,

$$k = k(s) = x'(s)y''(s) - x''(s)y'(s),$$

$$k = k(t) = \frac{\dot{x}(t)\ddot{y}(t) - \ddot{x}(t)\dot{y}(t)}{(\dot{x}^2(t) + \dot{y}^2(t))^{3/2}}$$

(the curvature of a planar curve may change its sign),
the Serret-Frénet formulas

$$\begin{pmatrix} \mathbf{T}'(s) \\ \mathbf{N}'(s) \end{pmatrix} = \begin{pmatrix} 0 & k(s) \\ -k(s) & 0 \end{pmatrix} \begin{pmatrix} \mathbf{T}(s) \\ \mathbf{N}(s) \end{pmatrix}.$$

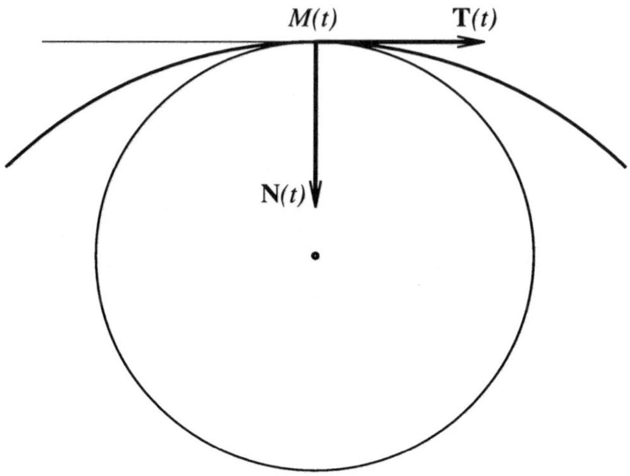

Figure 6.

B. Implicit description

A nonempty set γ of points $M(x, y)$, Cartesian coordinates x and y of which are connected by equality

$$F(x, y) = 0,$$

where $F(x, y)$ is a smooth function of variables x and y, is called *an implicitly defined planar curve* and the equality *an implicit equation* of the curve.

A point $M_0(x_0, y_0)$ is called *a regular point* of the implicitly given planar curve if the following conditions are met

$$F(x_0, y_0) = 0, \quad F_x^2(x_0, y_0) + F_y^2(x_0, y_0) > 0$$

and *a singular point* if

$$F(x_0, y_0) = 0, \quad F_x(x_0, y_0) = 0, \quad F_y(x_0, y_0) = 0.$$

3.1.7 Composite curves

Every so often a curve is constructed by assembling the specially selected curved fragments. For the resulting composite curve to have sufficiently good geometric characteristics, these fragments must be regular curves and in addition must satisfy certain requirements at the junction points.

Let

$$\mathbf{R} = \mathbf{R}_-(t), \quad a_- \leq t \leq b_-, \qquad \mathbf{R} = \mathbf{R}_+(t), \quad a_+ \leq t \leq b_+$$

be parametric equations of smooth curves γ_- and γ_+, respectively.

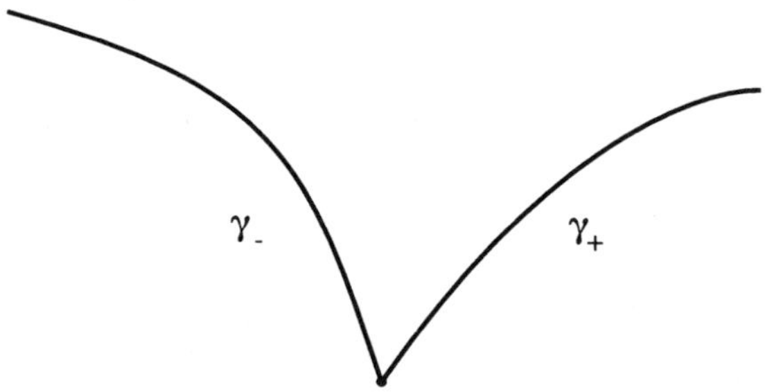

Figure 7.

Requirement 1

For a composite curve $\gamma_- \cup \gamma_+$ to be continuous, it is necessary that the right endpoint of curve γ_- coincides with the left endpoint of curve γ_+:

$$\mathbf{R}_-(b_-) = \mathbf{R}_+(a_+)$$

(Figure 7).

It is important to keep in mind that, on the one hand, a composite curve may possess outstanding geometric characteristics (as a set of points) but be defined by relatively poor parametric equations, and on the other hand, the curve may have the coordinate functions with high differential properties and be a nonregular curve.

Let us consider some examples.

Attention: examples

Example 1. A composite curve $\gamma = \gamma_- \cup \gamma_+$ is given by parametric equations

$$\mathbf{R}_-(t) = \begin{pmatrix} 4t \\ 4t \end{pmatrix}, \quad 0 \le t \le \frac{1}{4}, \quad \mathbf{R}_+(t) = \begin{pmatrix} 1+t \\ 1+t \end{pmatrix}, \quad 0 \le t \le 1$$

(Figure 8).

Result: The composite curve is a segment of a straight line; the tangent vectors at the common point are different,

$$\dot{\mathbf{R}}_-\left(\frac{1}{4}\right) = \begin{pmatrix} 4 \\ 4 \end{pmatrix}, \quad \dot{\mathbf{R}}_+(0) = \begin{pmatrix} 1 \\ 1 \end{pmatrix}.$$

Cause of singularity: the parametrization nonconformity at the joining point of the curves.

Requirement 2

$$\dot{\mathbf{R}}_+(a_+) = \alpha \dot{\mathbf{R}}_-(b_-), \quad \alpha > 0.$$

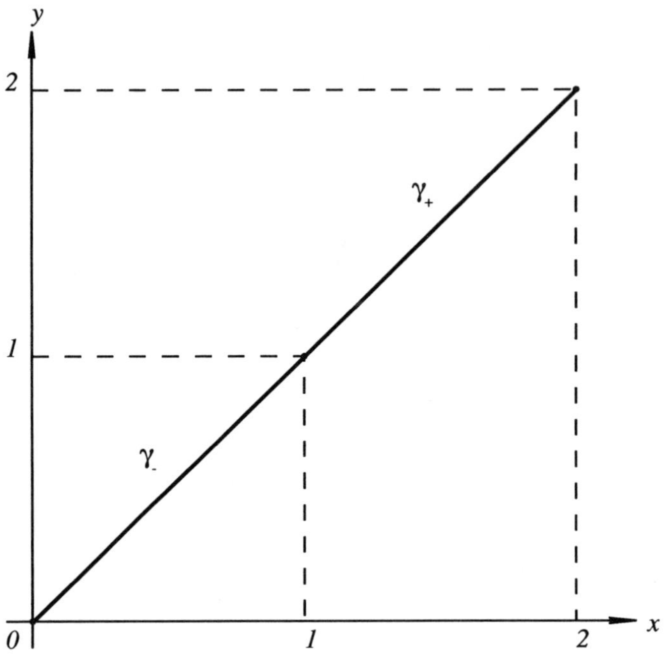

Figure 8.

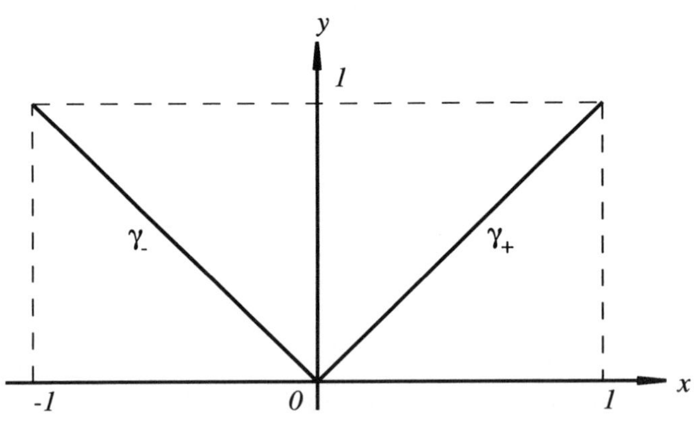

Figure 9.

Example 2. A composite curve $\gamma = \gamma_- \cup \gamma_+$ is defined by parametric equations

$$\mathbf{R}_-(t) = \left(\begin{array}{c} -(1-t)^2 \\ -(1-t)^2 \end{array} \right), \quad 0 \le t \le 1, \quad \mathbf{R}_+(t) = \left(\begin{array}{c} t^2 \\ t^2 \end{array} \right), \quad 0 \le t \le 1$$

(Figure 9).

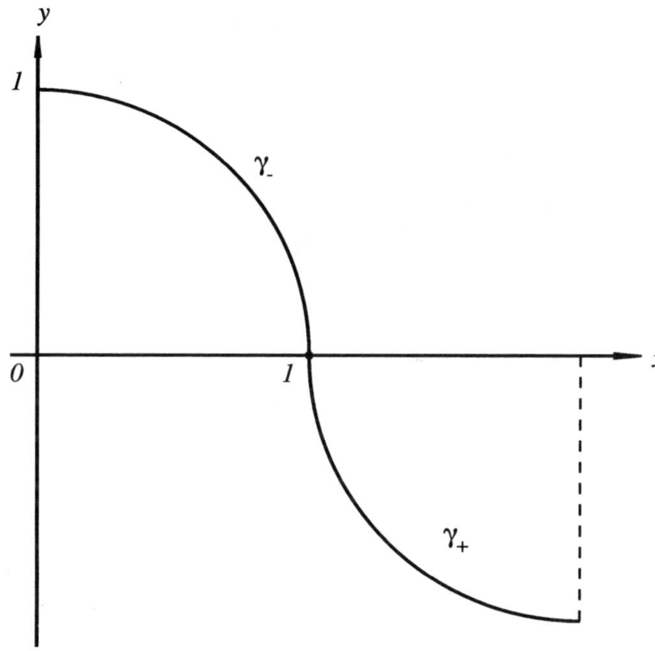

Figure 10.

Result: The composite curve has a fracture; the tangent vectors at the junction point are equal:

$$\dot{\mathbf{R}}_-(1) = \begin{pmatrix} 0 \\ 0 \end{pmatrix}, \quad \dot{\mathbf{R}}_+(0) = \begin{pmatrix} 0 \\ 0 \end{pmatrix}.$$

Cause of singularity: the unfortunate parametrization of the joining points — the tangent vectors at the joining point are equal to zero.

Example 3. A composite curve $\gamma = \gamma_- \cup \gamma_+$ is defined by parametric equations

$$\mathbf{R}_-(t) = \begin{pmatrix} \cos\left(\frac{\pi}{2}(1-t)^3\right) \\ \sin\left(\frac{\pi}{2}(1-t)^3\right) \end{pmatrix}, \quad 0 \le t \le 1,$$

$$\mathbf{R}_+(t) = \begin{pmatrix} 2 - \cos\left(\frac{\pi}{2}t^3\right) \\ -\sin\left(\frac{\pi}{2}t^3\right) \end{pmatrix}, \quad 0 \le t \le 1$$

(Figure 10).

Result: The composite curve is a circular two-link broken line, the tangent vectors at the joining point are equal to zero, the second derivatives of the radius-vectors at the joining point are equal to zero, and the curvature

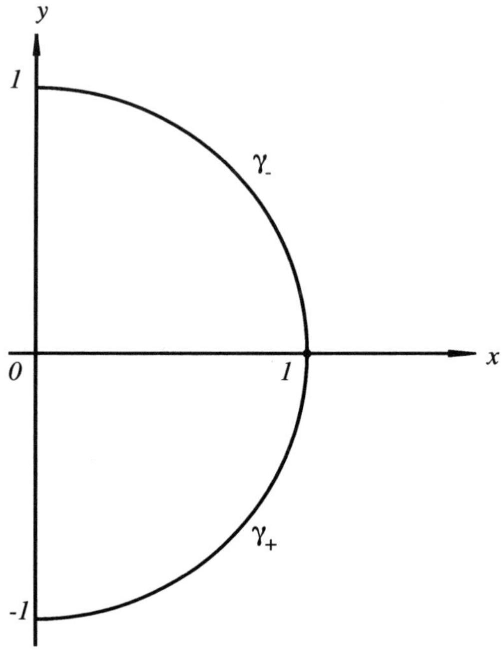

Figure 11.

vectors at the joining point are unequal

$$\mathbf{K}_{-}(1)=\begin{pmatrix}-1\\0\end{pmatrix},\quad \mathbf{K}_{+}(0)=\begin{pmatrix}1\\0\end{pmatrix}.$$

Cause of singularity: the parametrization unconformity of the joint curves at the joining point.

Example 4. A composite curve $\gamma = \gamma_{-} \cup \gamma_{+}$ is defined by parametric equations

$$\mathbf{R}_{-}(t)=\begin{pmatrix}\sin\left(\frac{\pi}{2}t^2\right)\\ \cos\left(\frac{\pi}{2}t^2\right)\end{pmatrix},\quad 0\le t\le 1,$$

$$\mathbf{R}_{+}(t)=\begin{pmatrix}\cos\left(\frac{\pi}{2}t^2\right)\\ -\sin\left(\frac{\pi}{2}t^2\right)\end{pmatrix},\quad 0\le t\le 1$$

(Figure 11).

Result: The composite curve is a circle; the tangent vectors are unequal,

$$\dot{\mathbf{R}}_{-}(1)=\begin{pmatrix}0\\-\frac{\pi}{2}\end{pmatrix},\quad \dot{\mathbf{R}}_{+}(0)=\begin{pmatrix}0\\0\end{pmatrix},$$

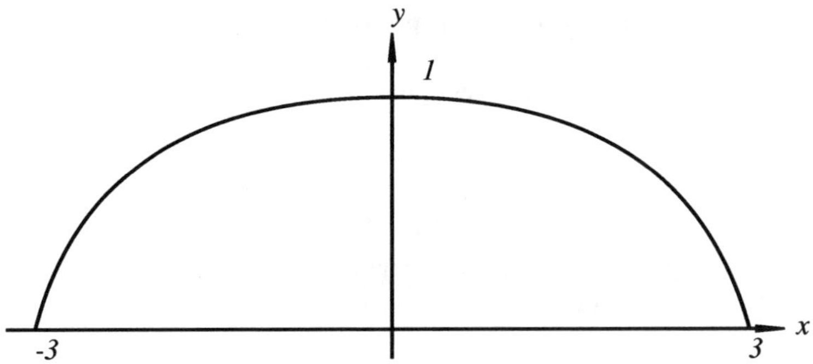

Figure 12.

the second derivatives of the radius-vectors at the joining point are unequal,

$$\ddot{\mathbf{R}}_-(1) = \begin{pmatrix} \pi^2 \\ -\pi \end{pmatrix}, \quad \ddot{\mathbf{R}}_+(0) = \begin{pmatrix} 0 \\ -\pi \end{pmatrix}$$

and the curvature vectors at the joining point are unequal,

$$\mathbf{K}_-(1) = \begin{pmatrix} -\frac{1}{a} \\ 0 \end{pmatrix}, \quad \mathbf{K}_+(0) = \begin{pmatrix} -\frac{1}{a} \\ 0 \end{pmatrix}.$$

Cause of singularity: one of the tangent vectors at the joining point is equal to zero.

Requirement 3

$$\dot{\mathbf{R}}_+(a_+) \neq \mathbf{O}, \quad \dot{\mathbf{R}}_-(b_-) \neq \mathbf{O}.$$

Example 5. A composite curve $\gamma = \gamma_- \cup \gamma_+$ is defined by parametric equations

$$\mathbf{R}_-(t) = \begin{pmatrix} 2t^2 + t - 3 \\ -t^2 + 2t \end{pmatrix}, \quad 0 \leq t \leq 1,$$

$$\mathbf{R}_+(t) = \begin{pmatrix} -2t^2 + 5t \\ -t^2 + 1 \end{pmatrix}, \quad 0 \leq t \leq 1$$

(Figure 12).

Result: The tangent vectors at the joining point are equal and nonzero,

$$\dot{\mathbf{R}}_-(1) = \begin{pmatrix} 5 \\ 0 \end{pmatrix}, \quad \dot{\mathbf{R}}_+(0) = \begin{pmatrix} 5 \\ 0 \end{pmatrix}$$

the second derivatives of the radius-vectors are nonzero and unequal, and

$$\ddot{\mathbf{R}}_-(1) = \begin{pmatrix} 4 \\ -2 \end{pmatrix}, \quad \ddot{\mathbf{R}}_+(0) = \begin{pmatrix} -4 \\ -2 \end{pmatrix}$$

the curvature vectors at the joining point are equal,

$$\mathbf{K}_-(1) = \begin{pmatrix} 0 \\ -2 \end{pmatrix},$$

$$\mathbf{K}_+(0) = \frac{1}{5\sqrt{5}} \begin{pmatrix} 0 \\ -2 \end{pmatrix}.$$

Cause of singularity: the parametrization nonconformity of the joint curves at the joining point.

3.1.8 Geometric continuity

Every so often one encounters a situation when each regular curve participating in the construction of a new composite curve has its own parametrization.

Important remark. The case when the first derivatives of the radius-vectors at the joining point are equal to zero is too difficult and it is excluded from our discussion.

The parametrization nonconformity accounts for many singularities at the joining points. The singularities are *singularities of parametrization.*

A continuous curve is called G^1*-continuous (geometrically continuous,* G^1*-curve)* if its tangent line is continuous along the curve.

Every G^1-continuous curve is C^1-regular with respect to the natural parametrization.

The G^1-continuous curve is described in Example 1 (Figure 8).

A G^1-continuous curve is called G^2*-continuous (geometrically continuous,* G^2*-curve)* if its curvature vector is continuous along the curve.

Every G^2-continuous curve is C^2-regular with respect to the natural parametrization.

The G^2-continuous curve is described in Example 5 (Figure 12).

Requirements

For C^2-regular curves

$$\gamma_- : \quad \mathbf{R} = \mathbf{R}_-(t), \quad a_- \le t \le b_-$$

and

$$\gamma_+ : \quad \mathbf{R} = \mathbf{R}_+(t), \quad a_+ \le t \le b_+$$

under which a composite curve $\gamma = \gamma_- \cup \gamma_+$ is G^2-continuous:

$$\mathbf{R}_+(a_+) = \mathbf{R}_-(b_-),$$

$$\dot{\mathbf{R}}_+(a_+) = \beta_1 \dot{\mathbf{R}}_-(b_-), \quad \beta_1 > 0,$$

$$\ddot{\mathbf{R}}_+(a_+) = \beta_1^2 \ddot{\mathbf{R}}_-(b_-) + \beta_2 \dot{\mathbf{R}}_-(b_-).$$

Any G^2-continuous curve is a curve which is C^2-regular with respect to the natural parametrization but is not necessarily C^2-regular with respect to other parametrization.

If a curve parameter is an arc length s, the following equalities

$$\mathbf{R}_+(s) = \mathbf{R}_-(s), \quad \mathbf{R}'_+(s) = \mathbf{R}'_-(s), \quad \mathbf{R}''_+(s) = \mathbf{R}''_-(s)$$

are fulfilled at every point of the G^2-continuous curve.

3.2 Bézier curves

3.2.1 Parametric equations of an elementary Bézier curve

For a given array of vertices

$$\mathbf{P} = \{\mathbf{P}_i(x_i, y_i, z_i), \quad i = 0, 1, \ldots, m\}$$

an (elementary) Bézier curve of degree m is defined by the vector equation

$$\mathbf{R}(t) = \sum_{i=0}^{m} B_i^m(t)\mathbf{P}_i, \quad 0 \le t \le 1,$$

where

$$B_i^m(t) = \binom{m}{i} t^i (1-t)^{m-i} = \frac{m!}{i!(m-i)!} t^i (1-t)^{m-i}$$

are *the Bernstein polynomials.*

A **matrix form** of the Bézier curve equation

$$\begin{pmatrix} x(t) \\ y(t) \\ z(t) \end{pmatrix} = \begin{pmatrix} x_0 & \ldots & x_m \\ y_0 & \ldots & y_m \\ z_0 & \ldots & z_m \end{pmatrix} \begin{pmatrix} \mu_{00} & \ldots & \mu_{0m} \\ \ldots & \ldots & \ldots \\ \mu_{m0} & \ldots & \mu_{mm} \end{pmatrix} \begin{pmatrix} t^0 \\ \vdots \\ t^m \end{pmatrix},$$

$$0 \le t \le 1,$$

where

$$\mu_{ij} = (-1)^{j-i} \binom{m}{j} \binom{j}{i}.$$

If the parameter domain is the line segment $a \le t \le b$, the equation of the Bézier curve may be written as follows:

$$\mathbf{R}(t) = \sum_{i=0}^{m} B_i^m \left(\frac{t-a}{b-a}\right) \mathbf{P}_i.$$

Important special case. When $m = 3$, we have an elementary cubic Bézier curve defined by four vertices and described by equation

$$\mathbf{R}(t) = (((1-t)\mathbf{P}_0 + 3t\mathbf{P}_1)(1-t) + 3t^2\mathbf{P}_2)(1-t) + t^3\mathbf{P}_3$$

$$0 \le t \le 1$$

or in the matrix form

$$\mathbf{R}(t) = \mathbf{PMT}, \quad 0 \le t \le 1,$$

where

$$\mathbf{R}(t) = \begin{pmatrix} x(t) \\ y(t) \\ z(t) \end{pmatrix}, \quad \mathbf{M} = \begin{pmatrix} 1 & -3 & 3 & 1 \\ 0 & 3 & -6 & 3 \\ 0 & 0 & 3 & -3 \\ 0 & 0 & 0 & 1 \end{pmatrix}, \quad \mathbf{T} = \begin{pmatrix} t^0 \\ t^1 \\ t^2 \\ t^3 \end{pmatrix},$$

$$\mathbf{P} = \begin{pmatrix} \mathbf{P}_0 & \mathbf{P}_1 & \mathbf{P}_2 & \mathbf{P}_3 \end{pmatrix} = \begin{pmatrix} x_0 & x_1 & x_2 & x_3 \\ y_0 & y_1 & y_2 & y_3 \\ z_0 & z_1 & z_2 & z_3 \end{pmatrix}.$$

Matrix $\mathbf{M}$ is called *a basic matrix of the cubic Bézier curve*, or simply *a Bézier basic matrix.*

3.2.2 Properties of the elementary Bézier curves

The Bernstein polynomials

$$B_i^m(t) = \binom{m}{i} t^i (1-t)^{m-i} = \frac{m!}{i!(m-i)!} t^i (1-t)^{m-i}$$

are used for the description of the elementary Bézier curve as the functional weighting coefficients. The polynomials have many interesting properties which are essential for investigating the Bézier curve behavior. Let us describe some of them.

The Bernstein polynomials

1^+ are nonnegative,

2^+ form a partition of unity

$$\sum_{i=0}^{m} B_i^m(t) = 1,$$

3^+ are independent of the vertices from array $\mathbf{P}$.

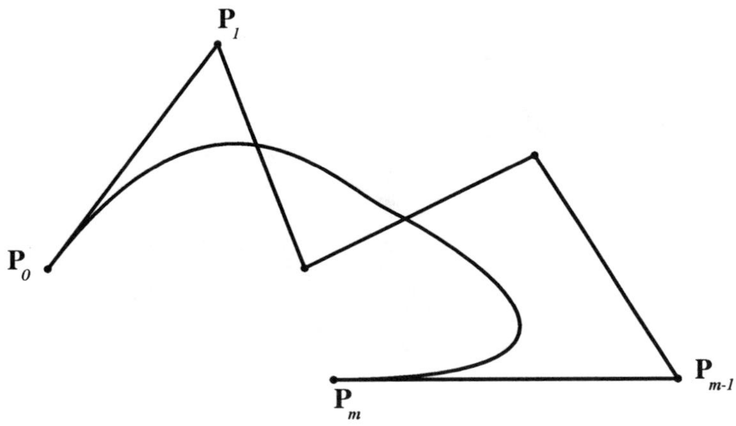

Figure 13.

Main properties of the Bézier curves

The elementary Bézier curve generated by array $\mathbf{P}$ possesses the following properties:

1^+ the curve is a smooth curve, in particular, the first and second derivatives of its radius-vector $\mathbf{R}(t)$ may be written in the form

$$\dot{\mathbf{R}}(t) = \frac{d}{dt}\mathbf{R}(t) = m \sum_{i=0}^{m-1} (\mathbf{P}_{i+1} - \mathbf{P}_i) B_i^{m-1}(t),$$

$$\ddot{\mathbf{R}}(t) = \frac{d^2}{dt^2}\mathbf{R}(t) = m(m-1) \sum_{i=0}^{m-2} (\mathbf{P}_{i+2} - 2\mathbf{P}_{i+1} + \mathbf{P}_i) B_i^{m-2}(t).$$

2^+ the curve starts at the first vertex $\mathbf{P}_0$ of array $\mathbf{P}$, $\mathbf{R}(0) = \mathbf{P}_0$, and its tangent at the vertex is parallel to segment $\mathbf{P}_0\mathbf{P}_1$ of the control polygon,

$$\dot{\mathbf{R}}(0) = m(\mathbf{P}_1 - \mathbf{P}_0)$$

and ends at the last vertex $\mathbf{P}_m$ of the array, $\mathbf{R}(1) = \mathbf{P}_m$, so that its tangent at the vertex is parallel to segment $\mathbf{P}_{m-1}\mathbf{P}_m$ of the control polygon,

$$\dot{\mathbf{R}}(1) = m(\mathbf{P}_m - \mathbf{P}_{m-1})$$

(Figure 13).

3^+ the curve lies in the convex hull generated by array $\mathbf{P}$ (Figure 14).

4^+ the curve is symmetric — the Bézier curve holds its form when the vertex order is changed to the opposite

$$\mathbf{P}_0, \mathbf{P}_1, \ldots, \mathbf{P}_{m-1}, \mathbf{P}_m \to \mathbf{P}_m, \mathbf{P}_{m-1}, \ldots, \mathbf{P}_1, \mathbf{P}_0.$$

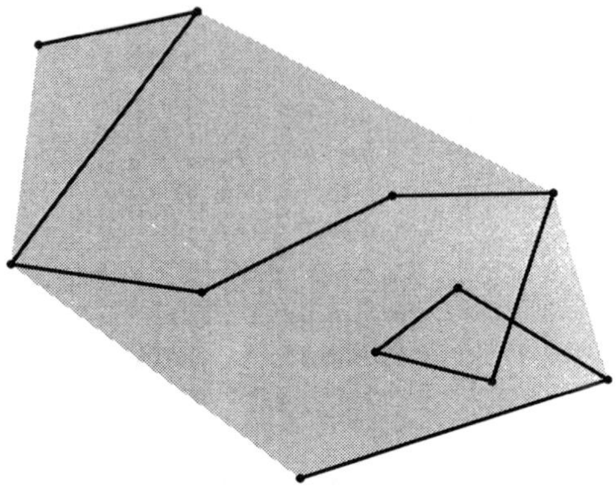

Figure 14.

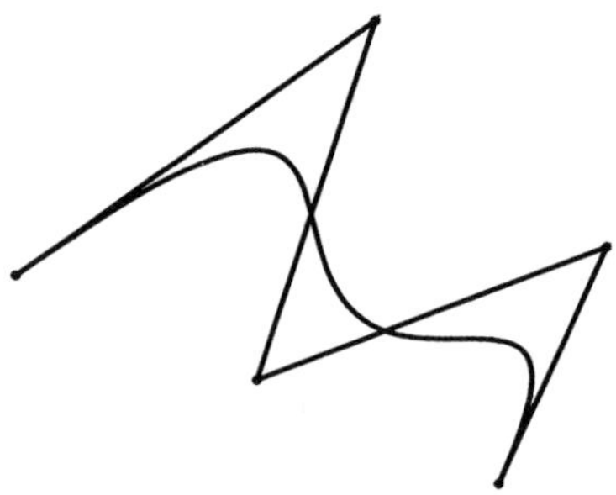

Figure 15a.

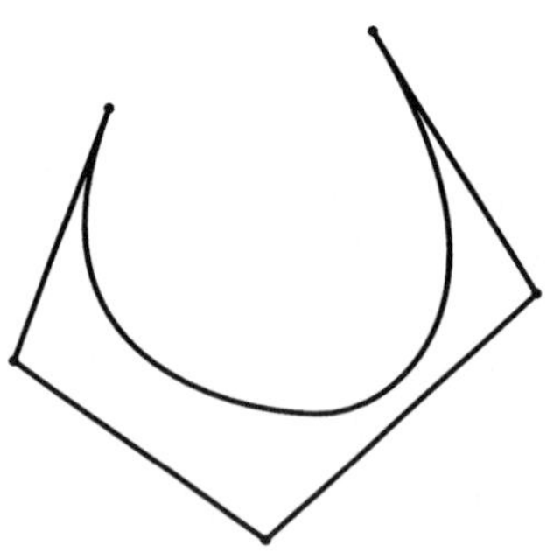

Figure 15b.

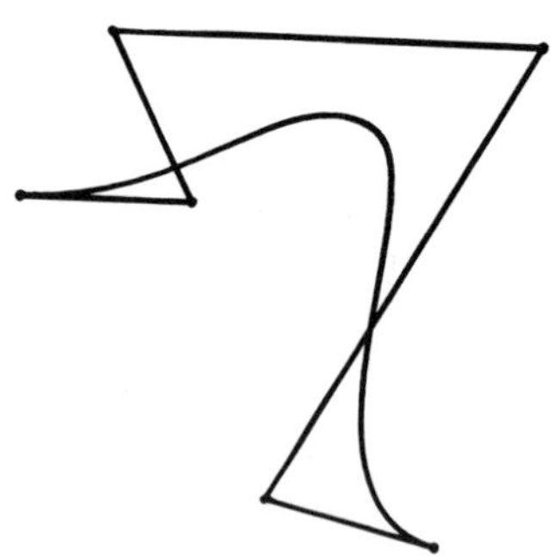

Figure 15c.

5^+ the curve is affine invariant.

6^+ the curve "mimicks" the control polygon (Figures 15a, 15b, and 15c) (in particular, the number of its intersection points with the arbitrary straight line is no more than the intersection number of the line and the control polygon).

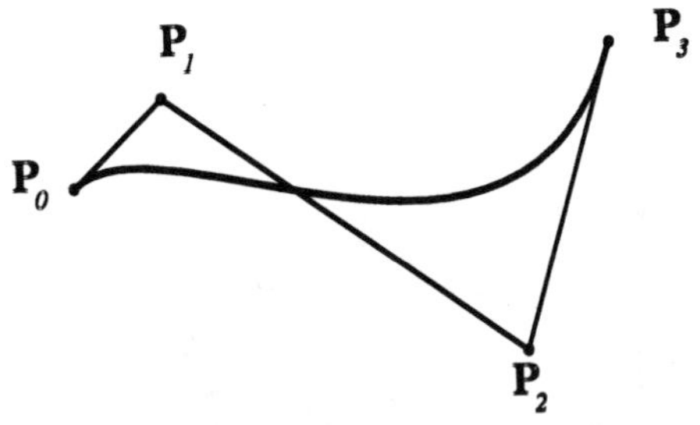

Figure 16a.

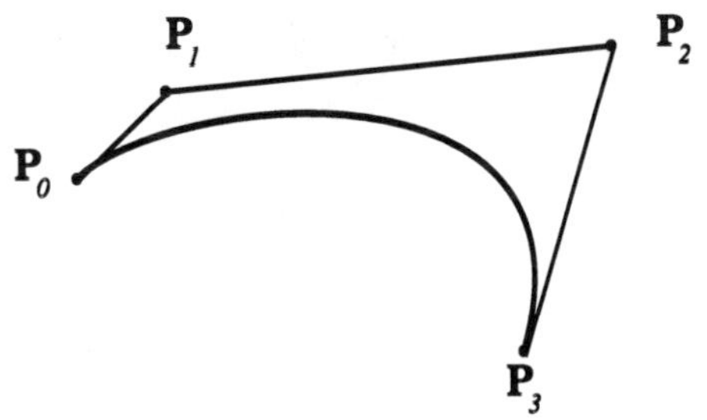

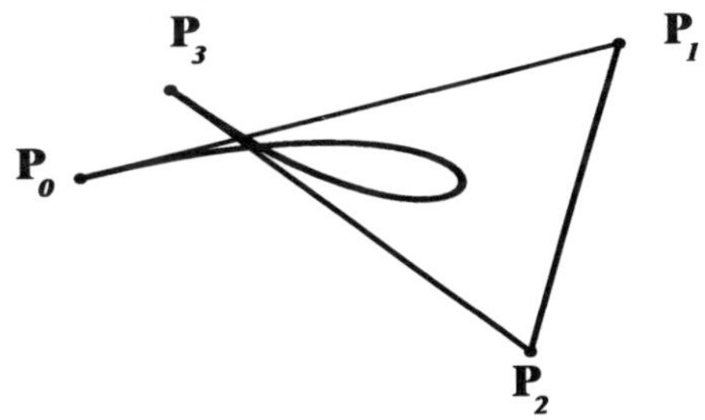

Figure 16b. **Figure 16c.**

7^+ the curve coincides with the line segment $\mathbf{P}_0\mathbf{P}_m$ if the support vertices $\mathbf{P}_0, \ldots, \mathbf{P}_m$ are collinear.

8^+ the curve lies in the plane if the support vertices $\mathbf{P}_0, \ldots, \mathbf{P}_m$ are coplanar.

9^- the degree of the Bernstein polynomials is directly related to the number of the control vertices and increases when it grows.

10^- with addition of only one vertex to the given array, it is necessary to recalculate the parametric equation of the elementary Bézier curve.

11^- the change of one vertex in the given array causes the whole curve shape to change.

12^- *a priori* information on arrangement of the Bézier curve (the convex hull of the given vertices contains the curve) is too rough (for a given quadruple of vertices in the plane (Figures 16a, 16b, and 16c) we have three various Bézier curves in its convex hull because the numeration order of vertices varies).

13^- the equation of the elementary Bézier curve does not contain free parameters (because of this, the curve is uniquely determined by the given array of vertices and there is no way of controlling its shape).

14^- the elementary Bézier curve is projective-noninvariant.

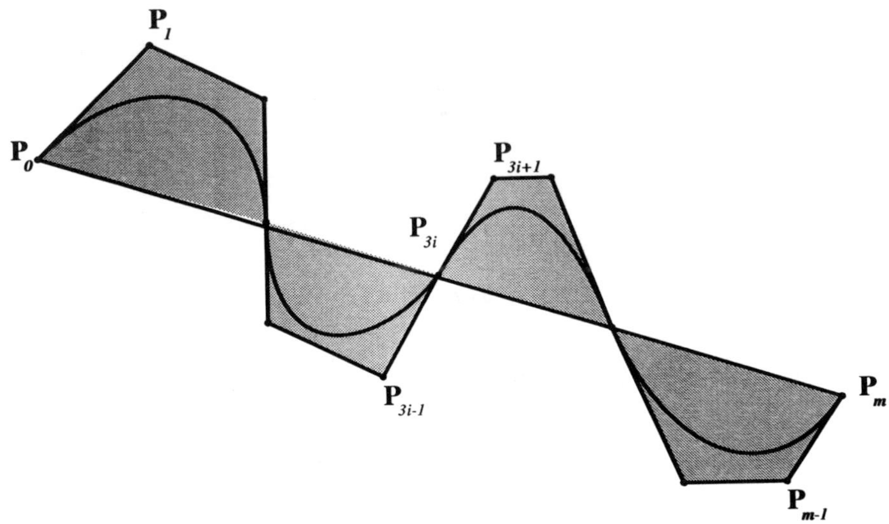

Figure 17.

3.2.3 Composite Bézier curves

In this section we consider the construction of the composite *cubic* Bézier curves in detail. Such curves are most often used in the applications. The composite Bézier curves of the other degrees are constructed in a similar way.

G^0-continuous curve γ is called *a composite cubic Bézier curve* if the curve is a union of the elementary Bézier curves $\gamma^{(1)}, \ldots, \gamma^{(l)}$,

$$\gamma = \gamma^{(1)} \cup \ldots \cup \gamma^{(l)}$$

so that the end point of the curve $\gamma^{(i)}$, $i = 0, \ldots, l-1$, coincides with the initial point of the curve $\gamma^{(i+1)}$.

If the curve $\gamma^{(i)}$ is specified by the parametric equation in the form

$$\mathbf{R} = \mathbf{R}^{(i)}(t), \quad 0 \le t \le 1$$

the condition may be written as follows:

$$\mathbf{R}^{(i)}(1) = \mathbf{R}^{(i+1)}(0), \quad i = 0, \ldots, l-1.$$

The smoothness conditions of composite cubic Bézier curve

In order for the composite cubic Bézier curve defined by the array of vertices $\mathbf{P}_0, \ldots, \mathbf{P}_m$ to be

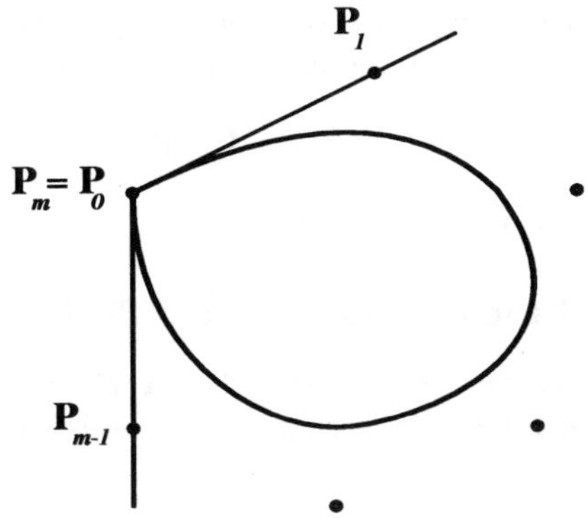

Figure 18a.

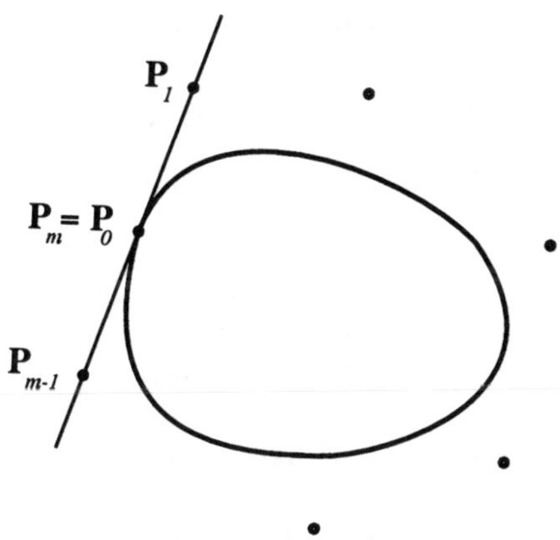

Figure 18b.

1) G^1-continuous curve, it is necessary for the triples of vertices

$$\mathbf{P}_{3i-1}, \mathbf{P}_{3i}, \mathbf{P}_{3i+1}, \quad (i \geq 1)$$

to be collinear (Figure 17),

2) G^2-continuous curve, it is necessary for the sets of five vertices

$$\mathbf{P}_{3i-2}, \mathbf{P}_{3i-1}, \mathbf{P}_{3i}, \mathbf{P}_{3i+1}, \mathbf{P}_{3i+2}, \quad (i \geq 1)$$

to be coplanar.

The closure conditions of the composite Bézier curve

In order for the composite cubic Bézier curve defined by the array of vertices

$$\mathbf{P}_0, \mathbf{P}_1, \ldots, \mathbf{P}_{m-1}, \mathbf{P}_m$$

to be a closed G^1-continuous composite Bézier curve, it is necessary for the vertices $\mathbf{P}_0$ and $\mathbf{P}_m$ to be coincident and for the triple

$$\mathbf{P}_{m-1}, \mathbf{P}_m = \mathbf{P}_0, \mathbf{P}_1$$

to be collinear (Figures 18a and 18b).

Parametrization of the composite cubic Bézier curves

When the composite curve γ is considered as a whole, it is more natural to use for its description the parametrization in the following form

$$\mathbf{R} = \mathbf{R}(t), \quad t_0 \le t \le t_l,$$

where

$$\mathbf{R} = \mathbf{R}^{(i)}(t), \quad t_{i-1} \le t \le t_i$$

is a parametric vector equation of the (i)-th elementary Bézier curve $\gamma^{(i)}$ $(i = 1, \dots, l)$.

The choice of the segment $[t_0, t_l]$ and the partial segments

$$[t_0, t_1], \dots, [t_{i-1}, t_i], \dots, [t_{n-1}, t_n]$$

on it merits special consideration.

The simplest way is to set $t_i = i$. This is called *uniform* or *equidistant* parametrization. When such (or equivalent) uniform subdivision of the parameter domain is too rigid, the other approaches may be used for choosing knots t_i.

Let us describe one method of parametrization for an array all three vertices of which

$$\mathbf{R}_{3i-1}, \mathbf{R}_{3i}, \mathbf{R}_{3i+1} \quad (i \ge 1)$$

are collinear. In this case the composite cubic Bézier curve has the shape shown in Figure 17 (a tangent line of the curve changes along the curve continuously). Let us choose a knot sequence t_i according to the rule:

$$t_0 = 0, \quad t_1 = 1, \quad \frac{t_i - t_{i-1}}{t_{i-1} - t_{i-2}} = \frac{|\,\mathbf{P}_{3i+1} - \mathbf{P}_{3i}\,|}{|\,\mathbf{P}_{3i} - \mathbf{P}_{3i-1}\,|}, \quad i = 2, \dots, l.$$

The curve γ is C^1-smooth with respect to the introduced parameter t and its tangent vector is continuous along the curve.

Remark. The curve shape is independent of the parametrization, but the parameter choice can influence its smoothness since the presence of the common tangent line at the joining point does not imply the continuity of the tangent vector. The smoothness of the curve depends not only on its shape, but on the choice of its parametrization as well.

If the given array of vertices does not contain the collinear triples, one can construct them by the addition of new vertices so that the composite curve defined by the new widened array is C^1-smooth.

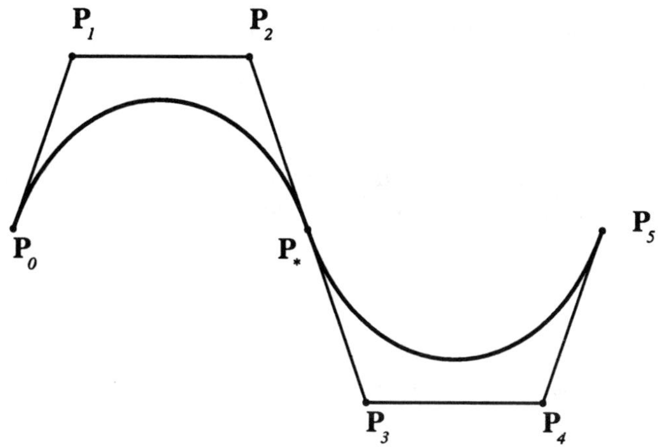

Figure 19.

Example. Let $\mathbf{P}$ be a sexuple of vertices

$$\mathbf{P}_0, \mathbf{P}_1, \mathbf{P}_2, \mathbf{P}_3, \mathbf{P}_4, \mathbf{P}_5.$$

Let us take, on the segment $\mathbf{P}_2\mathbf{P}_3$ of the control polygon, an auxiliary vertex $\mathbf{P}_*$ (for the sake of simplicity, a middle point of the segment $\mathbf{P}_2\mathbf{P}_3$) (Figure 19) and construct two elementary cubic Bézier curves for every quadruple of vertices $\mathbf{P}_0, \mathbf{P}_1, \mathbf{P}_2, \mathbf{P}_*$ and $\mathbf{P}_*, \mathbf{P}_3, \mathbf{P}_4, \mathbf{P}_5$.

$$\gamma^{(1)}: \ \mathbf{R} = \mathbf{R}^{(1)}(t), \quad 0 \le t \le 1,$$

$$\gamma^{(2)}: \ \mathbf{R} = \mathbf{R}^{(2)}(t), \quad 1 \le t \le 2.$$

The elementary fragments create a composite curve

$$\gamma: \ \mathbf{R} = \mathbf{R}(t) = \begin{cases} \mathbf{R}^{(1)}(t), & 0 \le t \le 1, \\ \mathbf{R}^{(2)}(t), & 1 \le t \le 2 \end{cases}$$

with a continuous tangent vector and a discontinuous curvature.

3.2.4 Rational Bézier curves

For a given array of vertices $\mathbf{P}_0, \ldots, \mathbf{P}_m$, *an (elementary) rational Bézier curve of degree m* is defined by an equation of the following form:

$$\mathbf{R}(t) = \frac{\sum_{i=0}^{m} w_i B_i^m(t) \mathbf{P}_i}{\sum_{i=0}^{m} w_i B_i^m(t)}, \quad 0 \le t \le 1,$$

where $B_i^m(t)$ are the Bernstein polynomials.

The nonnegative numbers w_i, the sum of which is positive, are called *(scalar) weights.* When all the weights w_i are equal, we have a standard elementary Bézier curve of the mth degree.

Properties of the rational Bézier curves

An elementary rational Bézier curve generated by array $\mathbf{P}$ possesses the following properties:

1^+ the curve is smooth.

2^+ the curve starts at the first vertex $\mathbf{P}_0$ of the array $\mathbf{P}$, $\mathbf{R}(0) = \mathbf{P}_0$, being tangent to the segment $\mathbf{P}_0\mathbf{P}_1$ of the control polygon,

$$\dot{\mathbf{R}}(0) = m\frac{w_1}{w_0}(\mathbf{P}_1 - \mathbf{P}_0)$$

and ends at the last vertex $\mathbf{P}_m$ of it, $\mathbf{R}(1) = \mathbf{P}_m$, being tangent to the segment $\mathbf{P}_{m-1}\mathbf{P}_m$ of the control polygon,

$$\dot{\mathbf{R}}(1) = m\frac{w_{m-1}}{w_m}(\mathbf{P}_m - \mathbf{P}_{m-1}).$$

3^+ the curve lies in the convex hull generated by the array of the support vertices $\mathbf{P}$.

4^+ the curve is symmetric; it holds its shape when the vertex order is reversed

$$\mathbf{P}_0, \mathbf{P}_1, \ldots, \mathbf{P}_{m-1}, \mathbf{P}_m \rightarrow \mathbf{P}_m, \mathbf{P}_{m-1}, \ldots, \mathbf{P}_1, \mathbf{P}_0.$$

5^+ the curve is affine invariant.

6^+ the curve "mimicks" the control polygon (in particular, the number of its intersection points with an arbitrary straight line is no more than the intersection number of the line and the control polygon).

7^+ the rational Bézier curve coincides with the segment $\mathbf{P}_0\mathbf{P}_m$ when the support vertices $\mathbf{P}_0, \ldots, \mathbf{P}_m$ are collinear.

8^+ if the support vertices $\mathbf{P}_0, \ldots, \mathbf{P}_m$ are coplanar, the rational Bézier curve is a planar curve.

9^+ the elementary rational Bézier curve is projectively invariant.

10^+ the behavior of the rational Bézier curve is determined by the array of vertices, as well as by the system of free parameters — the weights w_i, *the shape parameters*; for a given array of vertices the shape of the rational Bézier curve may be controlled using the weight factors.

11^- the degree of the functional coefficients is directly related to the vertex number and increases as the number grows.

12^- with addition of only one vertex to the given array, it is necessary to recalculate the parametric equation of the elementary Bézier curve.

13^- the change of one vertex in a given array causes the whole curve shape to change.

Property of the rational cubic Bézier curves

Let us consider two arrays $\mathbf{P}_0^{(1)}, \mathbf{P}_1^{(1)}, \mathbf{P}_2^{(1)}, \mathbf{P}_3^{(1)}$ and $\mathbf{P}_0^{(2)}, \mathbf{P}_1^{(2)}, \mathbf{P}_2^{(2)}, \mathbf{P}_3^{(2)}$. Under the condition that the vertices $\mathbf{P}_3^{(1)}$ and $\mathbf{P}_0^{(2)}$ coincide, the rational cubic Bézier curves $\gamma^{(1)}$ and $\gamma^{(2)}$ generated by the arrays have a common point.

Let us denote by $\gamma = \gamma^{(1)} \cup \gamma^{(2)}$ the obtained composite curve.

As may be seen from the relations

$$\dot{\mathbf{R}}^{(1)}(1) = 3\frac{w_2^{(1)}}{w_3^{(1)}}(\mathbf{P}_3^{(1)} - \mathbf{P}_2^{(1)}), \quad \dot{\mathbf{R}}^{(2)}(0) = 3\frac{w_1^{(2)}}{w_0^{(2)}}(\mathbf{P}_1^{(2)} - \mathbf{P}_0^{(2)})$$

under the conditions that the vertices

$$\mathbf{P}_2^{(1)}, \mathbf{P}_3^{(1)} = \mathbf{P}_0^{(2)}, \mathbf{P}_1^{(2)}$$

are collinear, we can choose the weights

$$w^{(1)} = (w_0^{(1)}, w_1^{(1)}, w_2^{(1)}, w_0^{(1)}), \qquad w^{(2)} = (w_0^{(2)}, w_1^{(2)}, w_2^{(2)}, w_3^{(2)})$$

$w^{(1)}$ and $w^{(2)}$ so that a tangent vector of the curve γ will be continuous along the curve. To do this it is sufficient to put

$$\frac{w_2^{(1)}}{w_3^{(1)}} \mid \mathbf{P}_3^{(1)} - \mathbf{P}_2^{(1)} \mid = \frac{w_1^{(2)}}{w_0^{(2)}} \mid \mathbf{P}_1^{(2)} - \mathbf{P}_0^{(2)} \mid .$$

If the vertices

$$\mathbf{P}_1^{(1)}, \mathbf{P}_2^{(1)}, \mathbf{P}_3^{(1)} = \mathbf{P}_0^{(2)}, \mathbf{P}_1^{(2)}, \mathbf{P}_2^{(2)}$$

are coplanar, that is, the vectors

$$(\mathbf{P}_2^{(1)} - \mathbf{P}_1^{(1)}) \wedge (\mathbf{P}_3^{(1)} - \mathbf{P}_2^{(1)}), \quad (\mathbf{P}_1^{(2)} - \mathbf{P}_0^{(2)}) \wedge (\mathbf{P}_2^{(2)} - \mathbf{P}_1^{(2)})$$

are collinear, we can choose the weights $w^{(1)}$ and $w^{(2)}$ so that a curvature vector of the curve γ will be continuous along the curve. To do this it is sufficient to choose them so that the following equality would be fulfilled:

$$\frac{W^{(1)}}{\mid \mathbf{P}_3^{(1)} - \mathbf{P}_2^{(1)} \mid^3} = \frac{W^{(2)}}{\mid \mathbf{P}_1^{(2)} - \mathbf{P}_0^{(2)} \mid^3},$$

where

$$W^{(i)} = \frac{w_{2-i}^{(i)} w_{4-i}^{(i)}}{(w_{3-i}^{(i)})^2} \mid (\mathbf{P}_{3-i}^{(i)} - \mathbf{P}_{2-i}^{(i)}) \wedge (\mathbf{P}_{4-i}^{(i)} - \mathbf{P}_{3-i}^{(i)}) \mid, \quad i = 1, 2.$$

The property of the rational cubic Bézier curves makes it possible to place an elementary cubic Bézier curve between any two given C^2-regular curves (in particular, between two elementary rational cubic Bézier curves) so that tangent and curvature vectors of the obtained composite curve will be continuous.

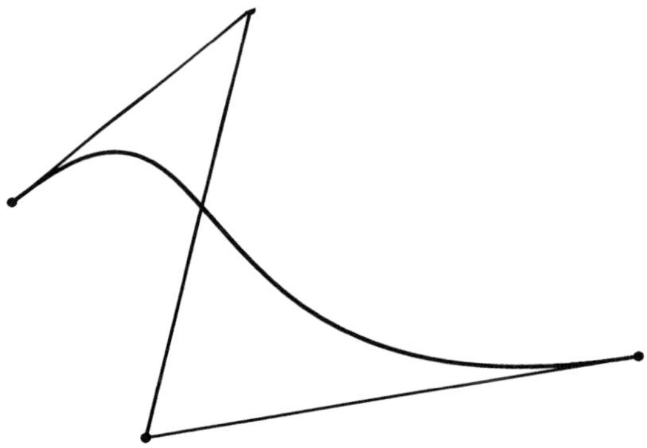

Figure 20.

Example. For a given quadruple of vertices $\mathbf{P}_0, \mathbf{P}_1, \mathbf{P}_2, \mathbf{P}_3$ and two non-negative numbers k_0 and k_3 it is required to find the weights w_0, w_1, w_2, w_3 and construct the rational cubic Bézier curve so that the curvature values at the end vertices of the control polygon $\mathbf{P}_0\mathbf{P}_1\mathbf{P}_2\mathbf{P}_3$ will coincide with the given numbers (with k_0 at vertex $\mathbf{P}_0$ and with k_3 at vertex $\mathbf{P}_3$).

Solution. The desired weights

$$w = (w_0, w_1, w_2, w_3)$$

are defined by the formulas

$$w_0 = 1, \quad w_1 = \frac{4}{3}\left(\frac{c_0^2 c_3}{k_0^2 k_3}\right)^{1/3}, \quad w_2 = \frac{4}{3}\left(\frac{c_0 c_3^2}{k_0 k_3^2}\right)^{1/3}, \quad w_3 = 1,$$

$$c_0 = \frac{|(\mathbf{P}_1 - \mathbf{P}_0) \wedge (\mathbf{P}_2) - \mathbf{P}_0)|}{2 \,|\, \mathbf{P}_1 - \mathbf{P}_0 \,|^3}, \quad c_3 = \frac{|(\mathbf{P}_2 - \mathbf{P}_1) \wedge (\mathbf{P}_3 - \mathbf{P}_1)|}{2 \,|\, \mathbf{P}_3 - \mathbf{P}_2 \,|^3}.$$

Remark. For a planar array $\mathbf{P}_0, \mathbf{P}_1, \mathbf{P}_2, \mathbf{P}_3$ with the control polygon of the S-shape (Figure 20), the numbers k_0 and k_1 must possess different signs (the curvature of the planar Bézier curve changes its sign).

3.2.5 Implementation

The following C function is an implementation of the algorithm described above.

Function syntax:

```
bezier( p, segments, div, output )
```

Function parameters are:

```
Flt p[]        (in)  - control polygon of Bezier curve
                       elements p[0] .. p[ m-1 ]
                       where m = 3*segments+1
int segments   (in)  - number of segments in Bezier curve
int div        (in)  - number of links in approximating
                       polygon
Flt output[]   (out) - approximating polygon
                       elements output[0] .. output[k-1]
                       where k = segments*div+1
```

The example below shows how to use this function to build a planar spline curve:

```
#include "lib.h"

/* Imported function declaration */
void bezier( Flt p[], int segments, int div, Flt output[] ) ;

/* number of points in control polygon */
/* M must be of form k*3+1 */
#define K    ( 2)
#define M    (K*3+1)
#define DIV  (10)
/* number of points in approximating polygon */
#define N    (K*DIV+1)

/* Control polygon initialization */
Flt xp[M] = { 0.295, 0.705, 0.266, 0.206,
              0.172, 0.508, 0.608 } ;
Flt yp[M] = { 0.154, 0.283, 0.413, 0.580,
              0.816, 0.745, 0.515 } ;
/* Declaration of output arrays */
Flt ox[N], oy[N] ;

void main( void )
{
    /* calculate spline */
    bezier( xp, K, DIV, ox ) ;
    bezier( yp, K, DIV, oy ) ;
    /* initialize graphics mode */
    SetVideoMode() ;
    /* display control polygon */
    DrawPolygon( xp, yp, M, 1 ) ;
    /* display built approximating polygon */
    DrawPolygon( ox, oy, N, 2 ) ;
}
```

The source of the program used may be found on the applied diskette in `bezier.c` file of the subdirectory `CURVES`. The detailed information on the file names and their contents is presented in `readme.txt` file of directory `SPLINES` and in Appendix B.

3.3 B-spline curves

3.3.1 Parametric equations of an elementary cubic B-spline curve

For a given array of vertices

$$\mathbf{P}_0, \mathbf{P}_1, \mathbf{P}_2, \mathbf{P}_3$$

(an elementary) cubic B-spline curve is defined by the vector equation of the following form:

$$\mathbf{R}(t) = \frac{(1-t)^3}{6}\mathbf{P}_0 + \frac{3t^3-6t^2+4}{6}\mathbf{P}_1 + \frac{-3t^3+3t^2+3t+1}{6}\mathbf{P}_2 + \frac{t^3}{6}\mathbf{P}_3,$$

$$0 \le t \le 1.$$

A **matrix form** of the parametric equation describing the elementary cubic B-spline curve:

$$\mathbf{R}(t) = \mathbf{PMT}, \quad 0 \le t \le 1,$$

where

$$\mathbf{R}(t) = \begin{pmatrix} x(t) \\ y(t) \\ z(t) \end{pmatrix}, \quad \mathbf{M} = \begin{pmatrix} 1 & -3 & 3 & -1 \\ 4 & 0 & -6 & 3 \\ 1 & 3 & 3 & -3 \\ 0 & 0 & 0 & 1 \end{pmatrix}, \quad \mathbf{T} = \begin{pmatrix} t^0 \\ t^1 \\ t^2 \\ t^3 \end{pmatrix},$$

$$\mathbf{P} = \begin{pmatrix} \mathbf{P}_0 & \mathbf{P}_1 & \mathbf{P}_2 & \mathbf{P}_3 \end{pmatrix} = \begin{pmatrix} x_0 & x_1 & x_2 & x_3 \\ y_0 & y_1 & y_2 & y_3 \\ z_0 & z_1 & z_2 & z_3 \end{pmatrix}$$

Matrix $\mathbf{M}$ is called *a basic matrix of cubic B-spline curve.*

Properties of the elementary cubic B-spline curve. Properties of the functional weight coefficients

$$n_0(t) = \frac{(1-t)^3}{6}, \qquad n_1(t) = \frac{3t^3-6t^2+4}{6},$$

$$n_2(t) = \frac{-3t^3+3t^2+3t+1}{6}, \qquad n_3(t) = \frac{t^3}{6}$$

are essential for the elementary cubic B-spline curve behavior. Let us describe some of them.

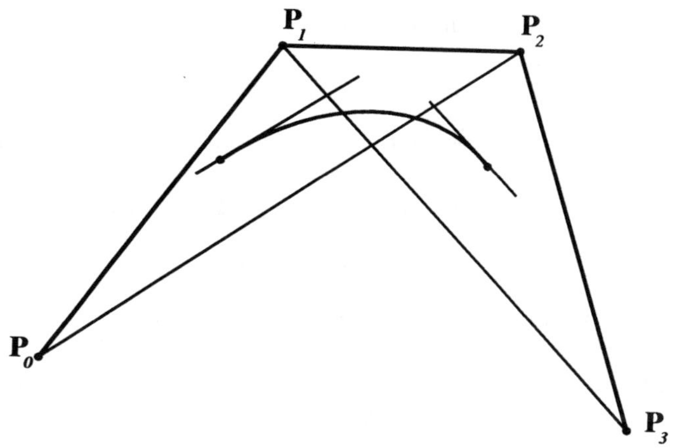

Figure 21.

The functional coefficients $n_i(t)$:
1^+ are nonnegative.
2^+ form a partition of unity

$$\sum_{i=0}^{3} n_i(t) = 1$$

3^+ are independent of the vertices $\mathbf{P}_0, \mathbf{P}_1, \mathbf{P}_2, \mathbf{P}_3$ (universal).

The elementary cubic B-spline curve generated by array $\mathbf{P}_0, \mathbf{P}_1, \mathbf{P}_2, \mathbf{P}_3$ possesses the following properties:

1^+ the curve lies in the convex hull generated by vertices $\mathbf{P}_0, \mathbf{P}_1, \mathbf{P}_2, \mathbf{P}_3$ of the control polygon and, as a rule, doesn't interpolate them.

2^+ its tangent line at the end point

$$\frac{1}{6}(\mathbf{P}_0 + 4\mathbf{P}_1 + \mathbf{P}_2)$$

is parallel to segment $\mathbf{P}_0\mathbf{P}_2$, and its tangent at the end point

$$\frac{1}{6}(\mathbf{P}_1 + 4\mathbf{P}_2 + \mathbf{P}_3)$$

is parallel to segment $\mathbf{P}_1\mathbf{P}_3$ (Figure 21).

3.3.2 Composite cubic B-spline curves

C^0-smooth (continuous) curve γ is called *a (composite) cubic B-spline curve* defined by array

$$\mathbf{P}_0, \ldots, \mathbf{P}_m, \quad m \geq 3$$

if the curve is a union of the elementary cubic B-spline curves $\gamma^{(1)}$, ..., $\gamma^{(m-2)}$,

$$\gamma = \gamma^{(1)} \cup \ldots \cup \gamma^{(m-2)},$$

where (i)th curve $\gamma^{(i)}$ is specified by the parametric equation of the form

$$\mathbf{R}^{(i)}(t) = \begin{pmatrix} \mathbf{P}_{i-1} & \mathbf{P}_i & \mathbf{P}_{i+1} & \mathbf{P}_{i+2} \end{pmatrix} \mathbf{MT},$$

$$0 \le t \le 1, \quad i = 1, \ldots, m-2,$$

where $\mathbf{M}$ is the basic matrix of the cubic B-spline curve.

Parametrization of the composite cubic B-spline curves. When the composite curve γ is considered as a whole, it is more natural to use a uniform parametrization for its description.

The simplest way is to choose *an equidistant* parametrization with the integer knots. For an array of $m+1$ support vertices, the composite B-spline curve is constructed from $m-2$ elementary fragments. If each of the fragments is defined on a unit segment, the common length of the parameter domain is equal to $m-2$. Let 0 be an initial point. In this case the parameter domain is segment $[0, m-2]$, and the parameter knots are defined by formulas

$$t_1 = 0, \quad t_{i+1} = t_i + 1, \quad i = 1, \ldots, m-3.$$

The choice of the parameter domain allows one to write the equation of the composite cubic B-spline curve γ in the following form:

$$\mathbf{R} = \mathbf{R}(t), \quad t_1 \le t \le t_m - 2,$$

where

$$\mathbf{R} = \mathbf{R}^{(i)}(t) = \begin{pmatrix} \mathbf{P}_{i-1} & \mathbf{P}_i & \mathbf{P}_{i+1} & \mathbf{P}_{i+2} \end{pmatrix} \mathbf{M} \begin{pmatrix} (t-t_i)^0 \\ (t-t_i)^1 \\ (t-t_i)^2 \\ (t-t_i)^3 \end{pmatrix},$$

$$t_i \le t \le t_{i+1}, \quad i = 1, \ldots, m-3$$

is a parametric vector equation of the (i)th elementary cubic B-spline curve $\gamma^{(i)}$.

Properties of the composite cubic B-spline curve. The composite cubic B-spline curve generated by array

$$\mathbf{P}_0, \ldots, \mathbf{P}_m, \quad m \ge 3$$

possesses the following properties:

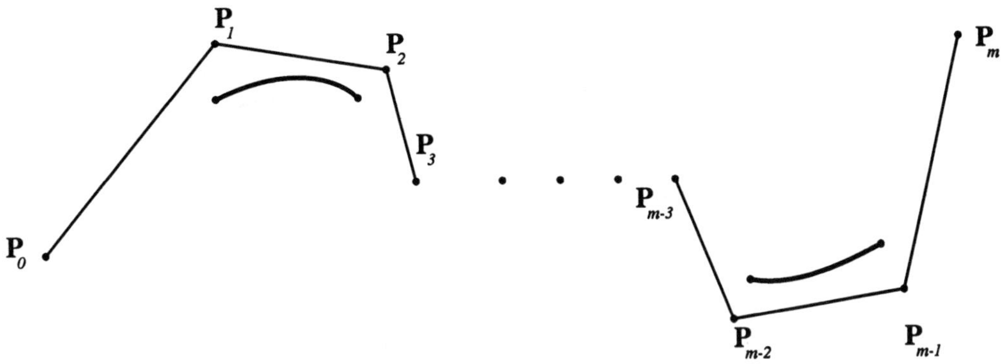

Figure 22.

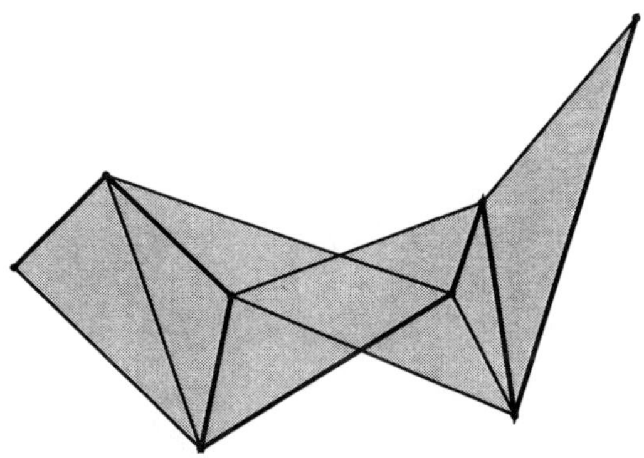

Figure 23.

1^+ the curve is a C^2-smooth curve ($t_{i+1} = t_i + 1, t_1 = 0$); the following equalities

$$\mathbf{R}^{(i)}(t_{i+1}) = \mathbf{R}^{(i+1)}(t_{i+1}) = \frac{1}{6}(\mathbf{P}_i + \mathbf{P}_{i+1} + \mathbf{P}_{i+2}),$$

$$\dot{\mathbf{R}}^{(i)}(t_{i+1}) = \dot{\mathbf{R}}^{(i+1)}(t_{i+1}) = \frac{1}{2}(-\mathbf{P}_i + \mathbf{P}_{i+2}),$$

$$\ddot{\mathbf{R}}^{(i)}(t_{i+1}) = \ddot{\mathbf{R}}^{(i+1)}(t_{i+1}) = \mathbf{P}_i - 2\mathbf{P}_{i+1} + \mathbf{P}_{i+2}$$

are true at the joining point of the elementary curves $\gamma^{(i)}$ and $\gamma^{(i+1)}$ (Figure 22).

2^+ as a rule, the curve doesn't interpolate the vertices of the given array.

3^+ the curve lies in the union of $m - 2$ convex hulls generated by the quadruples of vertices

$$\mathbf{P}_{i-1}, \mathbf{P}_i, \mathbf{P}_{i+1}, \mathbf{P}_{i+2}, \quad i = 1, \ldots, m-2$$

(Figure 23).

Figure 24.

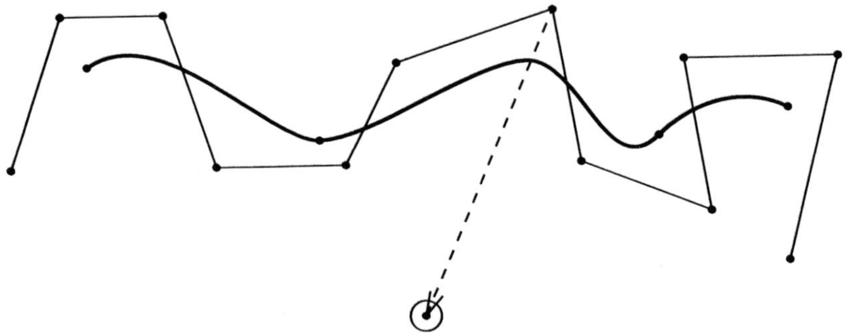

Figure 25a.

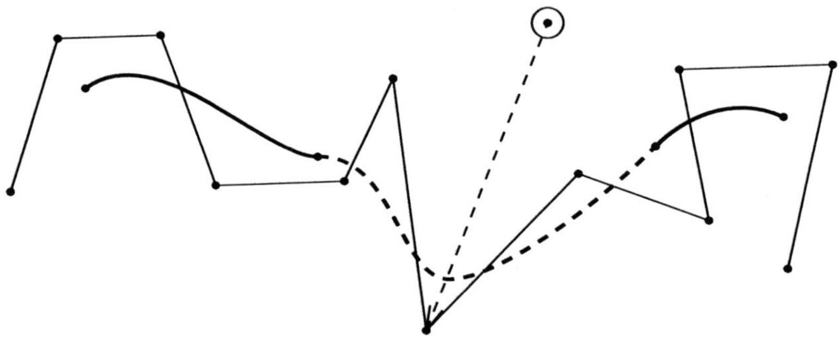

Figure 25b.

4^+ the curve "mimicks" the control polygon (Figure 24) (in particular, the number of its intersection points with the arbitrary straight line is no more than the intersection number of the line and the control polygon).

5^+ if the support vertices $\mathbf{P}_0, \ldots, \mathbf{P}_m$ of the array are collinear, then the composite cubic B-spline curve lies on the straight line (between vertices $\mathbf{P}_0$ and $\mathbf{P}_m$).

6^+ if the support vertices $\mathbf{P}_0, \ldots, \mathbf{P}_m$ of the array are coplanar, then the composite cubic B-spline curve lies in the plane containing these vertices.

7^+ the change of one vertex in the given array causes the curve to change in a region near the changed data only: when vertex $\mathbf{P}_i$ is changed we need to recalculate the parametric equations of four neighboring elementary curves $\gamma^{(i-2)}$, $\gamma^{(i-1)}$, $\gamma^{(i)}$, $\gamma^{(i+1)}$ (Figures 25a and 25b).

8^+ with addition of one vertex to the given array, it is necessary to recalculate the parametric equations of only four neighboring elementary curves in which description of the vertex takes part.

9^+ the composite cubic B-spline curve is affine invariant.

10^- the composite cubic B-spline curve is projective-noninvariant.

11^- the equations of the composite cubic B-spline curve do not contain free parameters (because of this, the curve is uniquely determined by the given array of vertices and there is no way of controlling its shape).

Remark. A mutual arrangement of the vertices in the array may be absolutely arbitrary (in particular, some of the vertices may coincide), but it should be remembered that in such a situation, the curve may lose its regularity. However, if the numbers of the coinciding vertices differ greatly, regularity is not necessarily lost.

A case when two or three first (last) vertices coincide is considered below.

3.3.3 Multiple and fantom vertices

As a rule, a composite cubic B-spline curve does not interpolate the vertices of the control array, but the arrangement of its end points is known: the initial point $\mathbf{R}^{(1)}(0)$ of the composite curve γ lies in triangle $\mathbf{P}_0\mathbf{P}_1\mathbf{P}_2$ and the final point $\mathbf{R}^{(m)}(1)$ lies in triangle $\mathbf{P}_{m-2}\mathbf{P}_{m-1}\mathbf{P}_m$. By choice of the auxiliary vertices and by construction of the additional elementary curves, one can gain an initial point of new composite curve which lies on segment $\mathbf{P}_0\mathbf{P}_1$, lies closer to vertex $\mathbf{P}_0$, or coincides with the vertex. Analogous results may be obtained for the other end point. For this purpose, it is common practice to use *multiple* or *fantom* vertices.

A. Double vertices. Let us assume that $\mathbf{P}_{-1} = \mathbf{P}_0$ and $\mathbf{P}_{m+1} = \mathbf{P}_m$ and construct a pair of new elementary curves $\gamma^{(0)}$ and $\gamma^{(m-1)}$ with the help of the parametric equations

$$\mathbf{R}^{(0)}(t) = (n_0(t) + n_1(t))\mathbf{P}_0 + n_2(t)\mathbf{P}_1 + n_3(t)\mathbf{P}_2,$$

$$\mathbf{R}^{(m-1)}(t) = n_0(t)\mathbf{P}_{m-2} + n_1(t)\mathbf{P}_{m-1} + (n_2(t) + n_3(t))\mathbf{P}_m,$$

where $0 \le t \le 1$. With curves $\gamma^{(0)}$ and $\gamma^{(m-1)}$, a new composite cubic B-spline curve

$$\gamma^* = \gamma^{(0)} \cup \gamma^{(1)} \cup \ldots \cup \gamma^{(m-2)} \cup \gamma^{(m-1)}$$

will start at point

$$\mathbf{R}^{(0)}(0) = \frac{5}{6}\mathbf{P}_0 + \frac{1}{6}\mathbf{P}_1$$

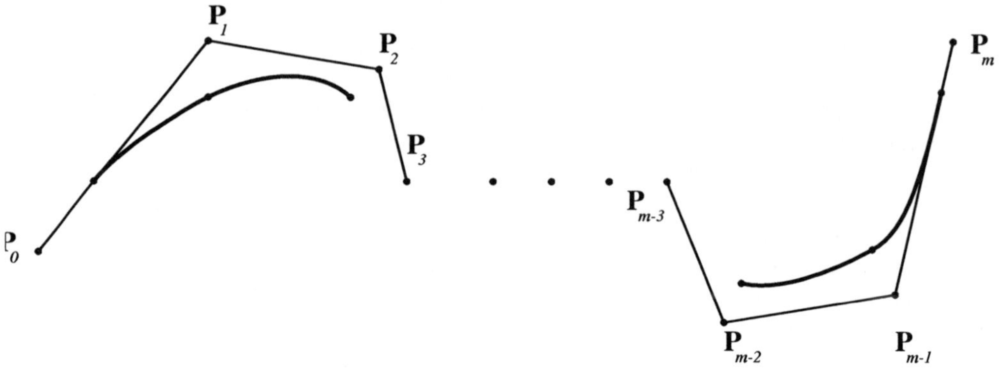

Figure 26a.

being tangent to segment $\mathbf{P}_0\mathbf{P}_1$,

$$\dot{\mathbf{R}}_2(0) = \frac{1}{2}(\mathbf{P}_1 - \mathbf{P}_0)$$

and finish at point

$$\mathbf{R}^{(m-1)}(1) = \frac{1}{6}\mathbf{P}_{m-1} + \frac{5}{6}\mathbf{P}_m$$

being tangent to segment $\mathbf{P}_{m-1}\mathbf{P}_m$,

$$\dot{\mathbf{R}}^{(m-1)}(1) = \frac{1}{2}(\mathbf{P}_m - \mathbf{P}_{m-1})$$

(Figure 26a). In addition, a curvature of curve γ^* is equal to zero at these end points.

B. Triple vertices. Let us assume that $\mathbf{P}_{-2} = \mathbf{P}_{-1} = \mathbf{P}_0$ and $\mathbf{P}_{m+2} = \mathbf{P}_{m+1} = \mathbf{P}_m$ and take straight-line segments

$$\mathbf{R}^{(-1)}(t) = \left(1 - \frac{t^3}{6}\right)\mathbf{P}_0 + \frac{t^3}{6}\mathbf{P}_1,$$

$$\mathbf{R}^{(m)}(t) = \frac{(1-t)^3}{6}\mathbf{P}_{m-1} + \left(1 - \frac{(1-t)^3}{6}\right)\mathbf{P}_m$$

where $0 \le t \le 1$ as two new elementary curves $\gamma^{(-1)}$ and $\gamma^{(m)}$(Figure 26b).

With curves $\gamma^{(0)}$ and $\gamma^{(m-1)}$, a new composite cubic B-spline curve

$$\gamma^{**} = \gamma^{(-1)} \cup \gamma^{(0)} \cup \gamma^{(1)} \cup \ldots \cup \gamma^{(m-2)} \cup \gamma^{(m-1)} \cup \gamma^{(m)}$$

will have vertices $\mathbf{P}_0$ and $\mathbf{P}_m$ as its end points.

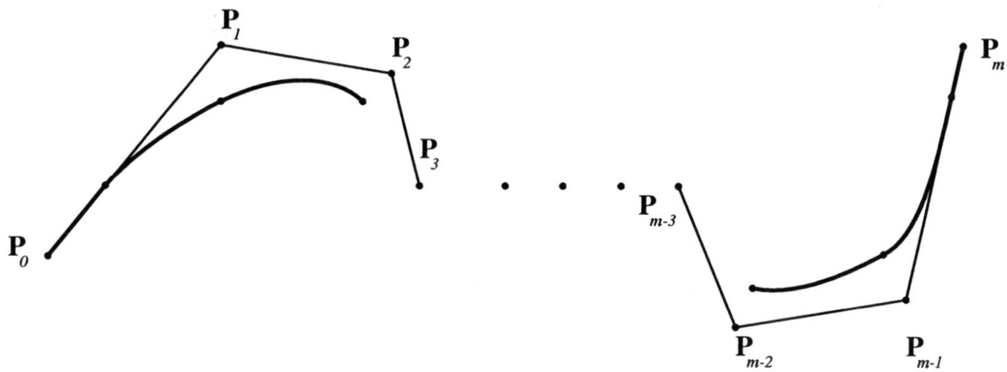

Figure 26b.

C. Fantom vertices. By choosing the additional vertices $\mathbf{P}_{-1}$ and $\mathbf{P}_{m+1}$ to array

$$\mathbf{P}_0, \ldots, \mathbf{P}_m, \quad m \geq 3$$

one can gain the fulfillment of the various conditions at the end points of the composite curve.

For example, a composite cubic B-spline curve constructed by new array

$$\mathbf{P}_{-1}, \mathbf{P}_0, \ldots, \mathbf{P}_m, \mathbf{P}_{m+1},$$

where

$$\mathbf{P}_{-1} = (\mathbf{P}_0 - \mathbf{P}_1) + \mathbf{P}_0, \quad \mathbf{P}_{m+1} = (\mathbf{P}_m - \mathbf{P}_{m-1}) + \mathbf{P}_m$$

will start at vertex $\mathbf{P}_0$ being tangent to segment $\mathbf{P}_0\mathbf{P}_1$,

$$\dot{\mathbf{R}}^{(0)}(0) = \mathbf{P}_1 - \mathbf{P}_0$$

and finish at vertex $\mathbf{P}_m$ being tangent to segment $\mathbf{P}_{m-1}\mathbf{P}_m$,

$$\dot{\mathbf{R}}^{(m-1)}(1) = \mathbf{P}_m - \mathbf{P}_{m-1}$$

(Figure 26c). In general, the curvature values of the new curve at points $\mathbf{P}_0$ and $\mathbf{P}_m$ are nonzero.

Remark. Additional vertices $\mathbf{P}_{-1}$ and $\mathbf{P}_{m+1}$ may be chosen such that the first and second derivatives of radius-vectors $\mathbf{R}^{(0)}(t)$ and $\mathbf{R}^{(m-1)}(t)$ of curves $\gamma^{(0)}$ and $\gamma^{(m-1)}$ at the end points of the new composite curve (for $t = 0$ and $t = 1$, respectively) will have the given values.

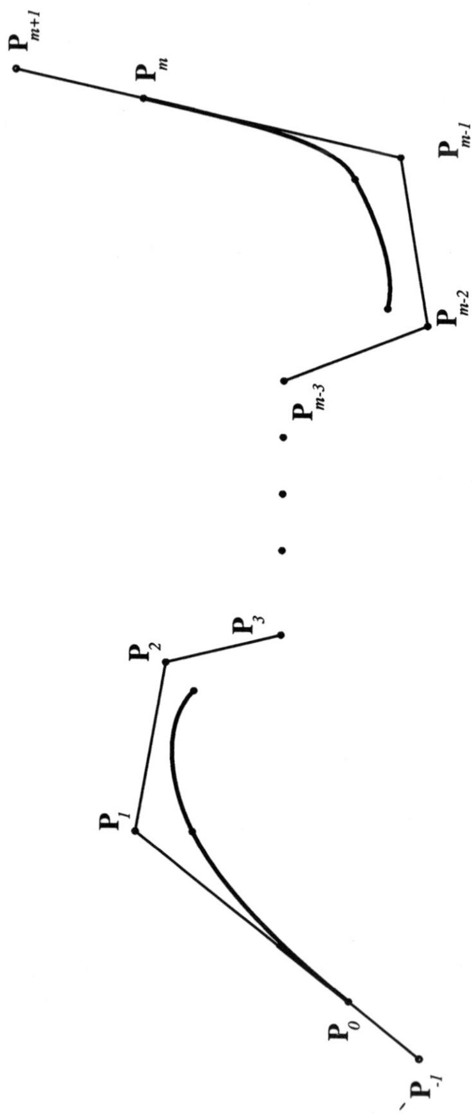
$\mathbf{P}_{m+1}$
$\mathbf{P}_m$
$\mathbf{P}_{m-1}$
$\mathbf{P}_{m-2}$
$\mathbf{P}_{m-3}$
$\mathbf{P}_3$
$\mathbf{P}_2$
$\mathbf{P}_1$
$\mathbf{P}_0$
$\mathbf{P}_{-1}$

Figure 26c.

Construction of the closed curve. To construct a C^2-smooth closed curve for the given array

$$\mathbf{P}_0, \ldots, \mathbf{P}_m, \quad m \geq 3$$

it is sufficient to choose the additional vertices $\mathbf{P}_{m+1}$, $\mathbf{P}_{m+2}$, $\mathbf{P}_{m+3}$ according to the rule

$$\mathbf{P}_{m+1} = \mathbf{P}_0, \mathbf{P}_{m+2} = \mathbf{P}_1, \mathbf{P}_{m+3} = \mathbf{P}_2$$

and to consider array

$$\mathbf{P}_0, \mathbf{P}_1, \mathbf{P}_2, \ldots, \mathbf{P}_m, \mathbf{P}_{m+1} = \mathbf{P}_0, \mathbf{P}_{m+2} = \mathbf{P}_1, \mathbf{P}_{m+3} = \mathbf{P}_2$$

(under condition $t_{i+1} = t_i + 1, t_1 = 0$).

Nonuniform disposition of knots. Let us consider a case when the distances between the neighboring knots t_i are unequal,

$$t_i - t_{i-1} \neq t_{i+1} - t_i.$$

For given array

$$\mathbf{P}_0, \ldots, \mathbf{P}_m, \quad m \geq 3$$

a composite cubic B-spline curve is defined in much the same way as above under the uniform subdivision of the parameter domain,

$$\mathbf{R}_i(t) = N_{i-1}^3(t)\mathbf{P}_{i-1} + N_i^3(t)\mathbf{P}_i + N_{i+1}^3(t)\mathbf{P}_{i+1} + N_{i+2}^3(t)\mathbf{P}_{i+2},$$

$$1 \leq i \leq m-2, \quad t_i \leq t \leq t_{i+1}$$

but in this case the functional coefficients essentially depend on the knot intervals and are defined recursively,

$$N_i^0(t) = \begin{cases} 1, & t_i \leq t < t_{i+1}, \\ 0, & t < t_i, t_{i+1} \leq t, \end{cases}$$

$$N_i^1(t) = \frac{t - t_i}{t_{i+1} - t_i} N_i^0(t) + \frac{t_{i+2} - t}{t_{i+2} - t_{i+1}} N_{i+1}^0(t),$$

$$N_i^2(t) = \frac{t - t_i}{t_{i+2} - t_i} N_i^1(t) + \frac{t_{i+3} - t}{t_{i+3} - t_{i+1}} N_{i+1}^1(t),$$

$$N_i^3(t) = \frac{t - t_i}{t_{i+3} - t_i} N_i^2(t) + \frac{t_{i+4} - t}{t_{i+4} - t_{i+1}} N_{i+1}^2(t).$$

Properties of the functional coefficients $N_i^3(t)$. The functional coefficients $N_i^3(t)$:

1^+ are nonnegative.

2^+ form a partition of unity

$$\sum_{i=0}^{m} N_i^3(t) = 1.$$

Remark. The choice of the parametrization may be absolutely arbitrary, but every so often the *chord length parametrization* is very convenient. Under the paramertrization the intervals between knots t_i are defined by the lengths of the corresponding chords

$$\begin{aligned} t_0 &= 0, \\ t_1 &= \mid \mathbf{P}_1 - \mathbf{P}_{-1} \mid, \\ t_i &= t_{i-1} + \mid \mathbf{P}_i - \mathbf{P}_{i-1} \mid, \quad i = 2, \ldots, m-1, \\ t_m &= t_{m-1} + \mid \mathbf{P}_{m+1} - \mathbf{P}_{m-1} \mid . \end{aligned}$$

(In the array $\mathbf{P}_{-1}, \mathbf{P}_0, \ldots, \mathbf{P}_m, \mathbf{P}_{m+1}$.)

3.3.4 Rational cubic B-spline curves

For a given array of vertices $\mathbf{P}_0, \ldots, \mathbf{P}_m$, *an (elementary) rational cubic B-spline curve* is defined by an equation of the following form:

$$\mathbf{R}(t) = \frac{\sum_{i=0}^{3} w_i n_i(t) \mathbf{P}_i}{\sum_{i=0}^{3} w_i b_i(t)}, \quad 0 \le t \le 1.$$

Nonnegative numbers w_i whose sum is positive are called *weights*. When all the weights w_i are the same we have a standard elementary cubic B-spline curve.

Properties of the composite rational cubic B-spline curves. The composite rational cubic B-spline curve generated by array P possesses the following properties:

1^+ the curve is a C^2-smooth curve (under conditions $t_{i+1} = t_i + 1$, $t_1 = 0$).

2^+ as a rule, the curve does not interpolate the vertices of the given array.

3^+ the curve lies in the union of $m-2$ convex hulls generated by the arrays of support vertices

$$\mathbf{P}_{i-3}, \mathbf{P}_{i-2}, \mathbf{P}_{i-1}, \mathbf{P}_i, \quad i = 3, \ldots, m.$$

4^+ the curve "mimicks" the control polygon (in particular, the number of its intersection points with the arbitrary straight line is no more than the intersection number of the line and the control polygon).

5^+ if vertices $\mathbf{P}_0, \ldots, \mathbf{P}_m$ are collinear, the composite rational cubic B-spline curve lies on the straight line (between vertices $\mathbf{P}_0$ and $\mathbf{P}_m$).

6^+ if vertices $\mathbf{P}_0, \ldots, \mathbf{P}_m$ are coplanar, the composite rational cubic B-spline curve lies in the plane containing these vertices.

7^+ the change of one vertex in the given array causes the curve to change in a region near the changed data only: when vertex $\mathbf{P}_i$ is changed we need to recalculate the parametric equations of four neighboring elementary curves $\gamma^{(i-2)}$, $\gamma^{(i-1)}$, $\gamma^{(i)}$. $\gamma^{(i+1)}$.

8^+ with addition of one vertex to the given array, it is necessary to recalculate the parametric equations of only four neighboring elementary curves in which the description of the vertex takes part.

9^+ the composite rational cubic B-spline curve is affine invariant.

10^+ the behavior of the composite rational cubic B-spline curve is controlled by the array of vertices as well as by the set of the free parameters — the weights w_i, *the shape parameters*; for a given array of vertices the shape of the composite rational cubic B-spline curve may be controlled using the weight factors.

11^+ the rational cubic B-spline curve is projectively invariant.

3.3.5 The Bézier form of the composite cubic B-spline curve

The same composite curve may be made up of the elementary cubic curves of the various classes, but it should be remembered that, for this purpose, it is necessary to take its own array of the control vertices for every class. Here we describe the way to recalculate the vertices of given array $\mathbf{P}$ to the vertices of new array $\mathbf{S}$ such that a composite cubic B-spline curve generated by array $\mathbf{P}$ and a composite cubic Bézier curve generated by array $\mathbf{S}$ will be geometrically indistinguishable. Using the Bézier curves is convenient because the Bernstein cubic polynomials $(1-t)^3$, $3t(1-t)^2$, $t^2(1-t)$, t^3 are more stable with respect to calculations in comparison with the monomial set 1, t, t^2, t^3.

Given vertices

$$\mathbf{P}_{-1}, \mathbf{P}_0, \ldots, \mathbf{P}_m, \mathbf{P}_{m+1}$$

and a sequence of knots $t_0 < \ldots < t_m$.

Let us assume that

$$\Delta_{i-1} = t_i - t_{i-1}, \quad i = 1, \ldots, m-1$$

and define vertices

$$\mathbf{S}_0, \mathbf{S}_1, \ldots, \mathbf{S}_{3m-1}, \mathbf{S}_{3m}$$

with formulas

$$\mathbf{S}_0 = \mathbf{P}_{-1}, \qquad \mathbf{S}_1 = \mathbf{P}_0,$$

$$\mathbf{S}_2 = \frac{\Delta_1}{\Delta_0 + \Delta_1}\mathbf{P}_0 + \frac{\Delta_0}{\Delta_0 + \Delta_1}\mathbf{P}_1,$$

$$\mathbf{S}_3 = \frac{\Delta_1}{\Delta_0 + \Delta_1}\mathbf{S}_2 + \frac{\Delta_0}{\Delta_0 + \Delta_1}\mathbf{S}_4,$$

$$\mathbf{S}_{3i-2} = \frac{\Delta_{i-1} + \Delta_i}{\Delta}\mathbf{P}_{i-1} + \frac{\Delta_{i-2}}{\Delta}\mathbf{P}_i,$$

$$\mathbf{S}_{3i-1} = \frac{\Delta_i}{\Delta}\mathbf{P}_{i-1} + \frac{\Delta_{i-2} + \Delta_{i-1}}{\Delta}\mathbf{P}_i,$$

$$\mathbf{S}_{3i} = \frac{\Delta_i}{\Delta_{i-1} + \Delta_i}\mathbf{S}_{3i-1} + \frac{\Delta_{i-1}}{\Delta_{i-1} + \Delta_i}\mathbf{S}_{3i+1},$$

$$\mathbf{S}_{3m-2} = \frac{\Delta_{m-1}}{\Delta_{m-2} + \Delta_{m-1}}\mathbf{P}_{m-1} + \frac{\Delta_{m-2}}{\Delta_{m-2} + \Delta_{m-1}}\mathbf{P}_m,$$

$$\mathbf{S}_{3m-1} = \mathbf{P}_m, \qquad \mathbf{S}_{3m} = \mathbf{P}_{m+1},$$

where

$$\Delta = \Delta_{i-2} + \Delta_{i-1} + \Delta_i, \quad i = 2, \ldots, m-1.$$

Definition. Polygon $\mathbf{P}_{-1} \ldots \mathbf{P}_{m+1}$ is called *a B-spline polygon*, polygon $\mathbf{S}_0 \ldots \mathbf{S}_{3m}$ is called a corresponding *Bézier polygon*, and vertices $\mathbf{S}_i$ are called *Bézier vertices.*

The C^2-smooth cubic B-spline curve defined by the array of vertices $\mathbf{P}_i$, $i = -1, \ldots, m+1$, coincides with the composite cubic Bézier curve defined by the array of the obtained vertices $\mathbf{S}_j$, $j = 0, \ldots, 3m$:

kth elementary curve of the composite Bézier curve is defined by the quadruple of vertices $\mathbf{S}_{3k+l}$, $(l = 0, 1, 2, 3,\ k = 0, \ldots, m-1)$.

The Bézier vertices for the rational cubic B-spline curves may be found in a similar way.

3.3.6 Implementation

The following C function is an implementation of the algorithm for constructing a cubic B-spline curve.

Function syntax:

```
bspline( p, m, div, output )
```

Function parameters are:

```
Flt p[]         (in)  - control polygon of spline
                        elements p[0] .. p[ m-1 ]
int m           (in)  - number of points in control polygon
int div         (in)  - number of links in approximating
                        polygon
Flt output[]    (out) - approximating polygon
                        elements output[0] .. [(m-3)*div]
```

The example below shows how to use this function to build a planar spline curve:

```
#include "lib.h"

/* Imported function declaration */
void bspline( Flt p[], int m, int div, Flt output[] ) ;

/* number of points in control polygon */
#define  M     (8)
#define DIV   (10)
/* number of points in approximating polygon */
#define  N    ((M-3)*DIV + 1)

/* Control polygon initialization */
Flt xp[M] = { 0.056, 0.287, 0.655, 0.716,
              0.228, 0.269, 0.666, 0.929 } ;
Flt yp[M] = { 0.820, 0.202, 0.202, 0.521,
              0.521, 0.820, 0.820, 0.227 } ;

/* Declaration of output arrays */
Flt ox[N], oy[N] ;

void main( void )
{
    /* calculate spline */
    bspline( xp, M, DIV, ox ) ;
    bspline( yp, M, DIV, oy ) ;
    /* initialize graphics mode */
    SetVideoMode() ;
    /* display control polygon */
    DrawPolygon( xp, yp, M, 1 ) ;
    /* display built approximating polygon */
    DrawPolygon( ox, oy, N, 2 ) ;
}
```

The source of the program used may be found on the applied diskette in `bspline.c` file of the subdirectory `CURVES`. The detailed information on the file names and their contents is presented in `readme.txt` file of directory `SPLINES` and in Appendix B.

The following C function is an implementation of the algorithm for constructing the nonuniform rational cubic B-splines (NURBS).

Function syntax:

```
nurbs( p, w, knot, m, div, output )
```

Function parameters are:

```
Flt p[]         (in)  - control polygon for NURBS
                        elements p[0] .. p[ m-1 ]
Flt w[]         (in)  - weights of points
                        elements w[0] .. w[ m-1 ]
Flt knot[]      (in)  - knots
                        elements knot[0] .. knot[ m+3 ]
int m           (in)  - number of points in control polygon
int div         (in)  - number of links in approximating
                        polygon
Flt output[]    (out) - approximating polygon of NURBS
                        elements output[0] .. [(m-3)*div]
```

The example below shows how to use this function to build a planar spline curve:

```
#include "lib.h"

/* Imported function declaration */
void nurbs( Flt p[], Flt w[], Flt knot[],
            int m, int div, Flt output[] ) ;

/* number of points in control polygon */
#define  M    (8)
#define DIV  (10)
/* number of points in approximating polygon */
#define  N   ((M-3)*DIV+1)

/* Control polygon initialization */
Flt xp[M] = { 0.056, 0.287, 0.655, 0.716,
              0.228, 0.269, 0.666, 0.929 } ;
Flt yp[M] = { 0.820, 0.202, 0.202, 0.521,
              0.521, 0.820, 0.820, 0.227 } ;
/* Weights of points */
Flt ww[M]={  1,   1,   1,   1,   7,   1,   1,   1 } ;
/* Knots */
Flt knot[M+4] = { 0, 1, 2, 3, 4, 5, 6, 7, 8, 9, 10, 11 } ;

/* Declaration of output arrays */
Flt ox[N], oy[N] ;

void main( void )
{
    /* calculate spline */
    nurbs( xp, ww, knot, M, DIV, ox ) ;
```

```
    nurbs( yp, ww, knot, M, DIV, oy ) ;
    /* initialize graphics mode */
    SetVideoMode() ;
    /* display control polygon */
    DrawPolygon( xp, yp, M, 1 ) ;
    /* display built approximating polygon */
    DrawPolygon( ox, oy, N, 2 ) ;
}
```

The source of the program used may be found on the applied diskette in `nurbs.c` file of the subdirectory `CURVES`. The detailed information on the file names and their contents is presented in `readme.txt` file of directory `SPLINES` and in Appendix B.

3.4 Beta-spline curves

3.4.1 Parametric equations of an elementary Beta-spline curve

For a given array of vertices

$$\mathbf{P}_0, \mathbf{P}_1, \mathbf{P}_2, \mathbf{P}_3$$

an (elementary) Beta-spline curve is defined by the vector equation of the following form:

$$\mathbf{R}(t) = b_0(t)\mathbf{P}_0 + b_1(t)\mathbf{P}_1 + b_2(t)\mathbf{P}_2 + b_3(t)\mathbf{P}_3, \quad 0 \le t \le 1,$$

in which the functional coefficients $b_i(t)$ are described as follows:

$$b_0(t) = \frac{2\beta_1^3}{\delta}(1-t)^3, \quad b_3(t) = \frac{2t^3}{\delta},$$

$$b_1(t) = \frac{1}{\delta}(2\beta_1^3 t(t^2 - 3t + 3) + 2\beta_1^2(t^3 - 3t^2 + 2) +$$
$$2\beta_1(t^3 - 3t + 2) + \beta_2(2t^3 - 3t^2 + 1)),$$

$$b_2(t) = \frac{1}{\delta}(2\beta_1^2 t^2(-t + 3) +$$
$$2\beta_1 t(-t^2 + 3) + \beta_2 t^2(-2t + 3) + 2(-t^3 + 1)),$$

where $\beta_1 > 0$, $\beta_2 \ge 0$, and $\delta = 2\beta_1^3 + 4\beta_1^2 + 4\beta_1 + \beta_2 + 2$.

The numerical parameters β_1 and β_2 are called *shape parameters* of the Beta-spline curve. In addition, the first parameter, β_1 is called *a bias parameter* and the second parameter, $\beta_2 \ge 0$, *a tension parameter.*

Remark. For $\beta_1 = 1$ and $\beta_2 = 0$ we have an elementary cubic B-spline curve.

A **matrix form** of the parametric equation describing the elementary Beta-spline curve

$$\mathbf{R}(t) = \mathbf{PMT}, \quad 0 \le t \le 1,$$

where

$$\mathbf{R}(t) = \begin{pmatrix} x(t) \\ y(t) \\ z(t) \end{pmatrix}, \qquad \mathbf{T} = \begin{pmatrix} t^0 \\ t^1 \\ t^2 \\ t^3 \end{pmatrix},$$

$$\mathbf{P} = \begin{pmatrix} \mathbf{P}_0 & \mathbf{P}_1 & \mathbf{P}_2 & \mathbf{P}_3 \end{pmatrix} = \begin{pmatrix} x_0 & x_1 & x_2 & x_3 \\ y_0 & y_1 & y_2 & y_3 \\ z_0 & z_1 & z_2 & z_3 \end{pmatrix},$$

$$\mathbf{M} = \frac{1}{\delta}\begin{pmatrix} 2\beta_1^3 & -6\beta_1^3 & 6\beta_1^3 & -2\beta_1^3 \\ 4(\beta_1^2+\beta_1)+\beta_2 & 6(\beta_1^3-\beta_1) & -3(2\beta_1^3+\eta) & 2(\beta_1^3+\theta) \\ 2 & 6\beta_1 & 3\eta & -2(\theta+1) \\ 0 & 0 & 0 & 2 \end{pmatrix}$$

(here $\eta = 2\beta_1^2 + \beta_2$, $\theta = \beta_1^2 + \beta_1 + \beta_2$).

Matrix $\mathbf{M}$ is called *a basic matrix of Beta-spline curve.*

Properties of the elementary Beta-spline curve. Properties of the functional weight coefficients $b_0(t), b_1(t), b_2(t), b_3(t)$ are essential for the elementary Beta-spline curve behavior. Let us describe some of them.

The functional weight coefficients $b_i(t)$:

1^+ are nonnegative.

2^+ form a partition of unity

$$\sum_{i=0}^{3} b_i(t) = 1$$

3^+ are independent of the vertices $\mathbf{P}_0$, $\mathbf{P}_1$, $\mathbf{P}_2$, $\mathbf{P}_3$ (universal).

The elementary Beta-spline curve generated by array $\mathbf{P}_0$, $\mathbf{P}_1$, $\mathbf{P}_2$, $\mathbf{P}_3$ possesses the following properties:

1^+ the curve lies in the convex hull generated by vertices $\mathbf{P}_0$, $\mathbf{P}_1$, $\mathbf{P}_2$, $\mathbf{P}_3$ of the control polygon (Figure 21) and, as a rule, does not interpolate them.

2^+ its tangent line at the end point

$$\frac{1}{\delta}(2\beta_1^3\mathbf{P}_0 + (4\beta_1^2 + 4\beta_1 + \beta_2)\mathbf{P}_1 + 2\mathbf{P}_2)$$

is parallel to vector

$$\frac{6}{\delta}\beta_1(\beta_1^2(\mathbf{P}_1 - \mathbf{P}_0) + (\mathbf{P}_2 - \mathbf{P}_1)$$

and at the end point

$$\frac{1}{\delta}(2\beta_1^3\mathbf{P}_1 + (4\beta_1^2 + 4\beta_1 + \beta_2)\mathbf{P}_2 + 2\mathbf{P}_3)$$

is parallel to vector

$$\frac{6}{\delta}(\beta_1^2(\mathbf{P}_2 - \mathbf{P}_1) + (\mathbf{P}_3 - \mathbf{P}_2).$$

3.4.2 Composite Beta-spline curves

C^0-smooth (continuous) curve γ is called *a (composite) Beta-spline curve* defined by array

$$\mathbf{P}_0, \ldots, \mathbf{P}_m, \quad m \geq 3,$$

if the curve is a union of $m-2$ elementary Beta-spline curves $\gamma^{(1)},\ldots,\gamma^{(m-2)}$,

$$\gamma = \gamma^{(1)} \cup \ldots \cup \gamma^{(m-2)}$$

where (i)th curve $\gamma^{(i)}$ is specified by the parametric equation of the form

$$\mathbf{R}^{(i)}(t) = \begin{pmatrix} \mathbf{P}_{i-1} & \mathbf{P}_i & \mathbf{P}_{i+1} & \mathbf{P}_{i+2} \end{pmatrix} \mathbf{MT}$$

$$0 \leq t \leq 1, \quad i = 1, \ldots, m-2$$

where $\mathbf{M}$ is the basic matrix of the Beta-spline curve.

Parametrization of the composite Beta-spline curve. When the composite curve γ is considered as a whole, it is more natural to use for its description a uniform parametrization.

The simplest way is to choose *an equidistant* parametrization with the integer knots. For an array of $m+1$ support vertices the composite Beta-spline curve is constructed from $m-2$ elementary fragments. If each of the fragments is defined on a unit segment, the common length of the parameter domain is equal to $m-2$. Let 0 be an initial point. In this case the parameter domain is segment $[0, m-2]$, and the parameter knots are defined by formulas

$$t_1 = 0, \quad t_{i+1} = t_i + 1, \quad i = 1, \ldots, m-3.$$

The described choice of the parameter domain allows one to write the equation of the composite Beta-spline curve γ in the following form:

$$\mathbf{R} = \mathbf{R}(t), \quad t_1 \leq t \leq t_{m-2},$$

where

$$\mathbf{R} = \mathbf{R}^{(i)}(t) = \begin{pmatrix} \mathbf{P}_{i-1} & \mathbf{P}_i & \mathbf{P}_{i+1} & \mathbf{P}_{i+2} \end{pmatrix} \mathbf{M} \begin{pmatrix} (t-t_i)^0 \\ (t-t_i)^1 \\ (t-t_i)^2 \\ (t-t_i)^3 \end{pmatrix},$$

$$t_i \le t \le t_{i+1}, \quad i = 1, \ldots, m-3$$

is the parametric vector equation of the (i)th elementary Beta-spline curve $\gamma^{(i)}$.

Properties of the composite Beta-spline curve. The composite Beta-spline curve generated by array

$$\mathbf{P}_0, \ldots, \mathbf{P}_m, \quad m \ge 3,$$

possesses the following properties:

1^+ the curve is a G^2-continuous curve; the following equalities

$$\mathbf{R}^{(i+1)}(t_{i+1}) = \mathbf{R}^{(i)}(t_{i+1}) = \frac{1}{\delta}(2\beta_1^3\mathbf{P}_i + (4\beta_1^2 + 4\beta_1 + \beta_2)\mathbf{P}_{i+1} + 2\mathbf{P}_{i+2}),$$

$$\dot{\mathbf{R}}^{(i+1)}(t_{i+1}) = \beta_1 \dot{\mathbf{R}}^{(i)}(t_{i+1}),$$

$$\ddot{\mathbf{R}}^{(i+1)}(t_{i+1}) = \beta_1^2 \ddot{\mathbf{R}}^{(i)}(t_{i+1}) + \beta_2 \dot{\mathbf{R}}^{(i)}(t_{i+})$$

are true at the joining point of the elementary curves $\gamma^{(i)}$ and $\gamma^{(i+1)}$.

2^+ as a rule, the curve does not interpolate the vertices of the given array.

3^+ the curve lies in the union of $m-2$ convex hulls generated by the quadruples of vertices

$$\mathbf{P}_{i-1}, \mathbf{P}_i, \mathbf{P}_{i+1}, \mathbf{P}_{i+2}, \quad i = 1, \ldots, m-2$$

(Figure 23).

4^+ the curve "mimicks" the control polygon (Figure 24) (in particular, the number of its intersection points with the arbitrary straight line is no more than the intersection number of the line and the control polygon).

5^+ if the support vertices $\mathbf{P}_0, \ldots, \mathbf{P}_m$ of the array are collinear, then the composite Beta-spline curve lies on the straight line (between vertices $\mathbf{P}_0$ and $\mathbf{P}_m$).

6^+ if the support vertices $\mathbf{P}_0, \ldots, \mathbf{P}_m$ of the array are coplanar, then the composite Beta-spline curve lies in the plane containing these vertices.

7^+ the change of one vertex in the given array causes the curve to change in a region near the changed data only: when vertex $\mathbf{P}_i$ is changed we need to recalculate the parametric equations of four neighboring elementary curves $\gamma^{(i-2)}$, $\gamma^{(i-1)}$, $\gamma^{(i)}$, $\gamma^{(i+1)}$ in construction of which the vertex takes part (Figures 25a and 25b).

8^+ with addition of one vertex to the given array, it is necessary to recalculate the parametric equations of only four elementary curves in which description of the vertex takes part.

9^+ the composite Beta-spline curve is affine invariant.

10^- the composite Beta-spline curve is projective-noninvariant.

11^+ the parameters β_1 and β_2 allow change of the shape of the (whole) composite Beta-spline curve.

Remark 1. The shape parameters β_1 and β_2 need not be the same for all of the elementary fragments. In view of the mutual arrangement of the vertices in the array, the parameters may be chosen in such a way that

1) a pair of numbers $\beta_1^{(i)}$ and $\beta_2^{(i)}$ is specific for every elementary fragment,

2) functions $\beta_1(t)$ and $\beta_2(t)$ depend on parameter t linearly on every unit segment [0,1], or

3) a pair of functions $\beta_1^{(i)}(t)$ and $\beta_2^{(i)}(t)$ $(i = 1, \ldots, m-2)$ is specific for each of the elementary fragments.

As a rule, the choice of the shape parameters β_1 and β_2 is defined by the mutual arrangement of the vertices in the array. If the distances between the neighboring vertices are approximately equal (not much different from each other), the choice of the parameters that are the same for all of the curve fragments gives sufficiently satisfactory results. If the mutual arrangement of the vertices in the array cannot be considered as uniform, good results are obtained by the suitable choice of the variable shape parameters.

Remark 2. A mutual arrangement of the vertices in the array may be absolutely arbitrary (in particular, some of the vertices may coincide), but it should be remembered that in such a situation, the curve may lose its regularity. However, if the numbers of the coinciding vertices differ greatly, a loss of regularity does not need to be.

A case when two or three first (last) vertices coincide is considered below.

3.4.3 Multiple and fantom vertices

As a rule, a composite Beta-spline curve does not interpolate the vertices of the control array that generated it, but the arrangement of its end points is known: the initial point $\mathbf{R}^{(1)}(0)$ of the composite curve γ lies in triangle $\Delta\mathbf{P}_0\mathbf{P}_1\mathbf{P}_2$ and the final point $\mathbf{R}^{(m-2)}(1)$ lies in triangle $\Delta\mathbf{P}_{m-2}\mathbf{P}_{m-1}\mathbf{P}_m$. By choice of the auxiliary vertices and by construction of the additional elementary curves one can gain an initial point of a new composite curve that lies on segment $\mathbf{P}_0\mathbf{P}_1$, lies closer to vertex $\mathbf{P}_0$, or coincides with the vertex. Analogous results may be obtained for the other end point. For this purpose, it is common practice to use *multiple* or *fantom* vertices.

A. Double vertices. Let us assume that $\mathbf{P}_{-1} = \mathbf{P}_0$ and $\mathbf{P}_{m+1} = \mathbf{P}_m$ and construct a pair of new elementary curves $\gamma^{(0)}$ and $\gamma^{(m-1)}$ with the help of the parametric equations

$$\mathbf{R}^{(0)}(t) = (b_0(t) + b_1(t)\mathbf{P}_0 + b_2(t)\mathbf{P}_1 + b_3(t)\mathbf{P}_2,$$

$$\mathbf{R}^{(m-1)}(t) = b_0(t)\mathbf{P}_{m-2} + b_1(t)\mathbf{P}_{m-1} + (b_2(t) + b_3(t))\mathbf{P}_m,$$

where $0 \le t \le 1$.

With curves $\gamma^{(0)}$ and $\gamma^{(m-1)}$, a new Beta-spline curve

$$\gamma^* = \gamma^{(0)} \cup \gamma^{(1)} \cup \ldots \cup \gamma^{(m-2)} \cup \gamma^{(m-1)}$$

will start at point

$$\mathbf{R}^{(0)}(0) = \left(1 - \frac{2}{\delta}\right)\mathbf{P}_0 + \frac{2}{\delta}\mathbf{P}_1$$

being tangent to segment $\mathbf{P}_0\mathbf{P}_1$,

$$\dot{\mathbf{R}}^{(0)}(0) = \frac{6\beta_1}{\delta}(\mathbf{P}_1 - \mathbf{P}_0)$$

and finish at point

$$\mathbf{R}^{(m-1)}(1) = \frac{2\beta_1^3}{\delta}\mathbf{P}_{m-1} + \left(1 - \frac{2\beta_1^3}{\delta}\right)\mathbf{P}_m$$

being tangent to segment $\mathbf{P}_{m-1}\mathbf{P}_m$,

$$\dot{\mathbf{R}}^{(m-1)}(1) = \frac{6\beta_1^2}{\delta}(\mathbf{P}_m - \mathbf{P}_{m-1})$$

(Figure 26a). In addition, a curvature of curve γ^* is equal to zero at these end points.

B. Triple vertices. Let us assume that $\mathbf{P}_{-2} = \mathbf{P}_{-1} = \mathbf{P}_0$ and $\mathbf{P}_{m+2} = \mathbf{P}_{m+1} = \mathbf{P}_m$ and take straight-line segments

$$\mathbf{R}^{(-1)}(t) = \left(1 - \frac{2t^3}{\delta}\right)\mathbf{P}_0 + \frac{2t^3}{\delta}\mathbf{P}_1,$$

$$\mathbf{R}^{(m)}(t) = \frac{2\beta_1^3(1-t)^3}{\delta}\mathbf{P}_{m-1}\left(1 - \frac{2\beta_1^3(1-t)^3}{\delta}\right)\mathbf{P}_m,$$

where $0 \le t \le 1$ as two new elementary curves $\gamma^{(-1)}$ and $\gamma^{(m)}$ (Figure 26b).

With curves $\gamma^{(0)}$ and $\gamma^{(m-1)}$ constructed previously, a new composite Beta spline curve

$$\gamma^{**} = \gamma^{(-1)} \cup \gamma^{(0)} \cup \gamma^{(1)} \cup \ldots \cup \gamma^{(m)} \cup \gamma^{(m-1)} \cup \gamma^m$$

will have vertices $\mathbf{P}_0$ and $\mathbf{P}_m$ as its end points.

C. Fantom vertices. By choosing the additional vertices $\mathbf{P}_{-1}$ and $\mathbf{P}_{m+1}$ to array

$$\mathbf{P}_0, \ldots, \mathbf{P}_m, \quad m \geq 3$$

one can gain the fulfillment of various conditions at the end points of the composite curve.

For example, a composite Beta-spline curve constructed by the new array

$$\mathbf{P}_{-1}, \mathbf{P}_0, \ldots, \mathbf{P}_m, \mathbf{P}_{m+1},$$

where

$$\mathbf{P}_{-1} = \frac{1}{\beta_1^3}(\mathbf{P}_0 - \mathbf{P}_1) + \mathbf{P}_0, \quad \mathbf{P}_{m+1} = \beta^3(\mathbf{P}_m - \mathbf{P}_{m-1}) + \mathbf{P}_m$$

will start at vertex $\mathbf{P}_0$ being tangent to segment $\mathbf{P}_0\mathbf{P}_1$,

$$\dot{\mathbf{R}}^{(0)}(0) = \frac{6(\beta_1 + 1)}{\delta}(\mathbf{P}_1 - \mathbf{P}_0)$$

and finish at vertex $\mathbf{P}_m$ being tangent to segment $\mathbf{P}_{m-1}\mathbf{P}_m$,

$$\dot{\mathbf{R}}^{(m-1)}(1) = \frac{6\beta_1^2(\beta_1 + 1)}{\delta}(\mathbf{P}_m - \mathbf{P}_{m-1})$$

(Figure 26c). In general, the curvature values of the new curve at points $\mathbf{P}_0$ and $\mathbf{P}_m$ are nonzero.

Remark. Additional vertices $\mathbf{P}_{-1}$ and $\mathbf{P}_{m+1}$ may be chosen such that the first and second derivatives of radius-vectors $\mathbf{R}^{(0)}(t)$ and $\mathbf{R}^{(m-1)}(t)$ of curves $\gamma^{(0)}$ and $\gamma^{(m-1)}$ at the end points of the new composite curve (for $t = 0$ and $t = 1$, respectively) will have the given values.

Construction of a closed curve. To construct G^2-continuous closed curve for given array

$$\mathbf{P}_0, \ldots, \mathbf{P}_m, \quad m \geq 3$$

it is sufficient to choose additional vertices $\mathbf{P}_{m+1}$, $\mathbf{P}_{m+2}$, $\mathbf{P}_{m+3}$ according to the rule

$$\mathbf{P}_{m+1} = \mathbf{P}_0, \mathbf{P}_{m+2} = \mathbf{P}_1, \mathbf{P}_{m+3} = \mathbf{P}_2$$

and then to consider array

$$\mathbf{P}_0, \mathbf{P}_1, \mathbf{P}_2, \ldots, \mathbf{P}_m, \mathbf{P}_{m+1} = \mathbf{P}_0, \mathbf{P}_{m+2} = \mathbf{P}_1, \mathbf{P}_{m+3} = \mathbf{P}_2.$$

3.4.4 Implementation

The following C function is an implementation of the algorithm described above.

Function syntax:

```
betaspline( b1, b2, p, m, div, output )
```

Function parameters are:

```
Flt b1, b2      (in)  - beta1 and beta2
Flt p[]         (in)  - control polygon of beta-spline,
                        elements p[0] .. p[m-1]
int m           (in)  - number of points in the control polygon
int div         (in)  - number of links in the approximating
                        polygon
Flt output[] (out) - approximating polygon of beta-spline,
                        elements
                        output[0] .. output [(m-3)*div]
```

The example below shows how to use this function to build a planar spline curve:

```
#include "lib.h"

/* Imported function declaration */
void betaspline( Flt b1, Flt b2, Flt p[],
                 int m, int div, Flt output[] ) ;

/* Number of points in control polygon       */
#define  M    (8)
#define DIV  (10)
/* Number of points in approximating polygon */
#define  N1   ((M-3)*DIV+1)
#define  N2   ((M-1)*DIV+1)
#define  N3   ((M+1)*DIV+1)

/* Control polygon initialization;           */
/* the first and the last points are tripled */
/* to demonstrate multiple vertices          */

Flt xp[M+4] = { 0.056, 0.056, 0.056, 0.287, 0.655, 0.716,
                0.228, 0.269, 0.666, 0.929, 0.929, 0.929 } ;
Flt yp[M+4] = { 0.820, 0.820, 0.820, 0.202, 0.202, 0.521,
                0.521, 0.820, 0.820, 0.227, 0.227, 0.227 } ;

Flt beta1 = 1.0, beta2 = 0.0 ;
```

```
/* Declaration of output arrays */
Flt ox1[N1], oy1[N1] ;
Flt ox2[N2], oy2[N2] ;
Flt ox3[N3], oy3[N3] ;

void main( void )
{
    /* calculate spline */
    betaspline( beta1, beta2, &xp[2], M, DIV, ox1 ) ;
    betaspline( beta1, beta2, &yp[2], M, DIV, oy1 ) ;

    betaspline( beta1, beta2, &xp[1], M+2, DIV, ox2 ) ;
    betaspline( beta1, beta2, &yp[1], M+2, DIV, oy2 ) ;

    betaspline( beta1, beta2, xp, M+4, DIV, ox3 ) ;
    betaspline( beta1, beta2, yp, M+4, DIV, oy3 ) ;

    /* initialize graphics mode */
    SetVideoMode() ;
    /* display control polygon */
    DrawPolygon( &xp[2], &yp[2], M, 1 ) ;
    /* display built approximating polygon */
    DrawPolygon( ox3, oy3, N3, 2 ) ;
    DrawPolygon( ox2, oy2, N2, 3 ) ;
    DrawPolygon( ox1, oy1, N1, 4 ) ;
}
```

The source of the program used may be found on the applied diskette in `beta.c` file of the subdirectory `CURVES`. The detailed information on the file names and their contents is presented in `readme.txt` file of directory `SPLINES` and in Appendix B.

3.5 Other spline curves

3.5.1 Interpolating cubic Hermite curves

For given vertices $\mathbf{P}_0$ and $\mathbf{P}_1$ and nonzero vectors $\mathbf{Q}_0$ and $\mathbf{Q}_1$, *an (elementary) cubic Hermite curve* is defined with the help of the vector equation

$$\mathbf{R}(t) = (1 - 3t^2 + 2t^3)\mathbf{P}_0 + t^2(3 - 2t)\mathbf{P}_1 + t(1 - 2t + t^2)\mathbf{Q}_0 - t^2(1 - t)\mathbf{Q}_1,$$
$$0 \le t \le 1.$$

A **matrix form** of the parametric equation described the elementary cubic Hermite curve:

$$\mathbf{R}(t) = \mathbf{GMT}, \quad 0 \le t \le 1,$$

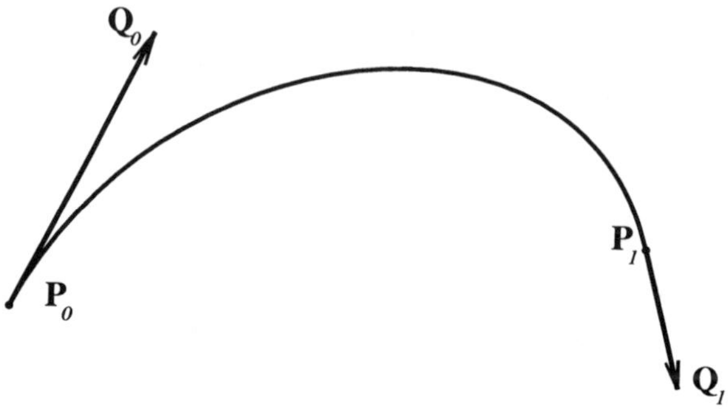

Figure 27.

where

$$\mathbf{G} = \begin{pmatrix} \mathbf{P}_0 & \mathbf{P}_1 & \mathbf{Q}_0 & \mathbf{Q}_1 \end{pmatrix} = \begin{pmatrix} x_0 & x_1 & u_0 & u_1 \\ y_0 & y_1 & v_0 & v_1 \\ z_0 & z_1 & w_0 & w_1 \end{pmatrix},$$

$$\mathbf{R} = \begin{pmatrix} x(t) \\ y(t) \\ z(t) \end{pmatrix}, \quad \mathbf{M} = \begin{pmatrix} 1 & 0 & -3 & 2 \\ 0 & 0 & 3 & -2 \\ 0 & 1 & -2 & 1 \\ 0 & 0 & -1 & 1 \end{pmatrix}, \quad \mathbf{T} = \begin{pmatrix} t^0 \\ t^1 \\ t^2 \\ t^3 \end{pmatrix}.$$

Matrix $\mathbf{M}$ is called *a basic matrix of cubic Hermite curve*, and matrix $\mathbf{G}$ its *geometric matrix*.

A tangent vector of the elementary cubic Hermite curve at the end point $\mathbf{P}_0$, $\mathbf{R}(0) = \mathbf{P}_0$, coincides with the given vector $\mathbf{Q}_0$, $\dot{\mathbf{R}}(0) = \mathbf{Q}_0$, and at the end point $\mathbf{P}_1$, $\mathbf{R}(1) = \mathbf{P}_1$, the tangent vector coincides with vector $\mathbf{Q}_1$, $\dot{\mathbf{R}}(1) = \mathbf{Q}_1$ (Figure 27).

Under an affine transformation of the parameter

$$t \mapsto \tau = (1-t)a + tb,$$

when a unit segment [0,1] goes into segment $[a, b]$, the shape of the elementary cubic Hermite curve changes. To hold the shape it is necessary to replace vectors $\mathbf{Q}_0$ and $\mathbf{Q}_1$ with vectors

$$\frac{1}{b-a}\mathbf{Q}_0 \quad \text{and} \quad \frac{1}{b-a}\mathbf{Q}_1,$$

respectively.

In addition to the above equations of the elementary cubic Hermite curve there exist some other forms of its description.

Equation of the elementary cubic Hermite curve

In the Bézier form

$$\mathbf{R}(t) = B_0^3(t)\mathbf{P}_0 + B_1^3(t)(\mathbf{P}_0 + \frac{1}{3}\mathbf{Q}_0) + B_2^3(t)(\mathbf{P}_1 - \frac{1}{3}\mathbf{Q}_1) + B_3^3(t)\mathbf{P}_1,$$

where $B_i^3(t)$ are the cubic polynomials of Bernstein, and

in the Hermite form

$$\mathbf{R}(t) = H_0^3(t)\mathbf{P}_0 + H_1^3(t)\mathbf{Q}_0) + H_2^3(t)\mathbf{Q}_1 + H_3^3(t)\mathbf{P}_1,$$

where $H_i^3(t)$ are the cubic polynomials of Hermite,

$$\begin{array}{rclrcl} H_0^3(t) & = & B_0^3(t) + B_1^3(t), & H_1^3(t) & = & \frac{1}{3}B_1^3(t), \\ H_2^3(t) & = & -\frac{1}{3}B_2^3(t), & H_3^3(t) & = & B_2^3(t) + B_3^3(t). \end{array}$$

Remark. The description of the Hermite curve in the Bernstein form is preferable because of the higher stability of the Bernstein polynomials.

Composite cubic Hermite curves. Curve γ is called *a (composite) cubic Hermite curve* defined by array

$$\mathbf{P}_0, \ldots, \mathbf{P}_m, \quad m \geq 1$$

and a pair of nonzero vectors $\mathbf{Q}_0$ and $\mathbf{Q}_m$ if the curve may be represented as a union of the elementary cubic Hermite curves $\gamma^{(1)}$, ..., $\gamma^{(m)}$,

$$\gamma = \gamma^{(1)} \cup \ldots \cup \gamma^{(m)};$$

(i)th curve $\gamma^{(i)}$ is described by the parametric equation of the form

$$\mathbf{R}^{(i)}(t) = \begin{pmatrix} \mathbf{P}_{i-1} & \mathbf{P}_i & \mathbf{Q}_{i-1} & \mathbf{Q}_i \end{pmatrix} \mathbf{MT}$$

$$0 \leq t \leq 1, \quad i = 1, \ldots, m,$$

where $\mathbf{M}$ is the basic matrix of the Hermite curve and vectors $\mathbf{Q}_1$, ..., $\mathbf{Q}_{m-1}$ are defined by the matrix equation of the following form:

$$\begin{pmatrix} 1 & 4 & 1 & & & & \\ & 1 & 4 & 1 & & & \\ & & & \ddots & & & \\ & & & & 1 & 4 & 1 \\ & & & & & 1 & 4 & 1 \end{pmatrix} \begin{pmatrix} \mathbf{Q}_0 \\ \mathbf{Q}_1 \\ \vdots \\ \mathbf{Q}_{m-1} \\ \mathbf{Q}_m \end{pmatrix}$$

$$= \begin{pmatrix} -3 & 0 & 3 & & & & \\ & -3 & 0 & 3 & & & \\ & & & \ddots & & & \\ & & & & -3 & 0 & 3 \\ & & & & & -3 & 0 & 3 \end{pmatrix} \begin{pmatrix} \mathbf{P}_0 \\ \mathbf{P}_1 \\ \vdots \\ \mathbf{P}_{m-1} \\ \mathbf{P}_m \end{pmatrix}.$$

Parametrization of the composite Hermite curve. When the composite curve γ is considered as a whole, it is more natural to use a single parametrization for its description.

The simplest parametrization is a uniform parametrization with the integer knots. For an array of $m+1$ support vertices a composite cubic Hermite curve is constructed from m elementary fragments. If each of them is defined on a unit segment, then a length of a common parameter segment must be equal to m. Let us take 0 as an initial point of the parameter domain. Then the domain is segment $[0, m]$. In this case, the parameter knots are defined by formulas

$$t_0 = 0, \quad t_i = t_{i-1} + 1, \quad i = 1, \ldots, m.$$

The choice of the parameter domain allows one to write the equation of the composite cubic Hermite curve γ as follows:

$$\mathbf{R} = \mathbf{R}(t), \quad t_0 \le t \le t_m,$$

where

$$\mathbf{R} = \mathbf{R}^{(i)}(t) = \begin{pmatrix} \mathbf{P}_{i-1} & \mathbf{P}_i & \mathbf{Q}_{i-1} & \mathbf{Q}_i \end{pmatrix} \mathbf{M} \begin{pmatrix} (t-t_i)^0 \\ (t-t_i)^1 \\ (t-t_i)^2 \\ (t-t_i)^3 \end{pmatrix},$$

$$t_{i-1} \le t \le t_i, \quad i = 1, \ldots, m$$

is a parametric vector equation of the (i)th elementary cubic Hermite curve $\gamma^{(i)}$.

Properties of the composite cubic Hermite curve. A composite cubic Hermite curve generated by array

$$\mathbf{P}_0, \ldots, \mathbf{P}_m, \quad m \ge 1$$

and a pair of nonzero vectors $\mathbf{Q}_0$ and $\mathbf{Q}_m$ possesses the following properties:

1^+ the curve is a C^2-smooth curve (the curve with continuous curvature).

2^+ the curve interpolates vertices $\mathbf{P}_0$, ..., $\mathbf{P}_m$ and at the joining point of the elementary curves $\gamma^{(i)}$ and $\gamma^{(i+1)}$ we have

$$\mathbf{R}^{(i)}(t_i) = \mathbf{R}^{(i+1)}(t_i) = \mathbf{P}_i$$

(Figure 28).

3^+ a tangent vector $\mathbf{Q}_i$ at the inner vertex $\mathbf{P}_i$ ($i = 1, \ldots, m-1$) is uniquely determined by the vertices of array $\mathbf{P}$ and the tangent vectors $\mathbf{Q}_0$ and $\mathbf{Q}_m$ at the end points $\mathbf{P}_0$ and $\mathbf{P}_m$, respectively.

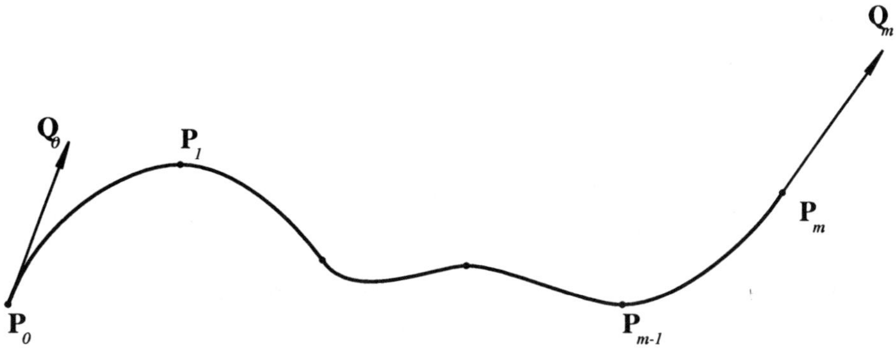

Figure 28.

4^- the curve does not lie in the convex hull generated by the given array.

5^- the change of one vertex in a given array or one of the tangent vectors at the end points causes the whole curve shape to change.

6^- with addition of only one vertex to the given array, it is necessary to recalculate all the parametric equations.

7^+ the composite cubic Hermite curve is affine invariant.

8^- the equations of the composite cubic Hermite curve do not contain free parameters (because of this, the curve is uniquely determined by the given array of vertices and the pair of the tangent vectors and there is no way of controlling its shape).

9^- the cubic Hermite curve is projectively noninvariant.

Remark. Using the elementary cubic Hermite curves, a composite curve may be constructed by given array

$$\mathbf{P}_0, \ldots, \mathbf{P}_m, \quad m \geq 1$$

and an arbitrary set of nonzero vectors

$$\mathbf{Q}_0, \ldots, \mathbf{Q}_m.$$

Every quadruple $\mathbf{P}_{i-1}$, $\mathbf{P}_i$, $\mathbf{Q}_{i-1}$, $\mathbf{Q}_i$ defines an elementary cubic Hermite curve. The curve is described by the parametric equation of the form

$$\mathbf{R} = \mathbf{R}^{(i)}(t) = \begin{pmatrix} \mathbf{P}_{i-1} & \mathbf{P}_i & \mathbf{Q}_{i-1} & \mathbf{Q}_i \end{pmatrix} \mathbf{M} \begin{pmatrix} (t-t_i)^0 \\ (t-t_i)^1 \\ (t-t_i)^2 \\ (t-t_i)^3 \end{pmatrix},$$

$$t_{i-1} \leq t \leq t_i, \quad i = 1, \ldots, m.$$

But, contrary to the above-described situation, in this case the composite curve is only C^1-smooth.

Implementation

The following C function is an implementation of the algorithm described above.

Function syntax:

```
hermite( p, m, q0, qm, div, output )
```

Function parameters are:

```
Flt p[]        (in)  - control polygon of spline
                       elements p[0] .. p[ m-1 ]
int m          (in)  - number of points in control polygon
Flt q0, qm     (in)  - corresponding components of
                       tangent vectors
int div        (in)  - number of links in approximating
                       polygon
Flt output[] (out) - approximating polygon of hermite
                       elements output[0] .. [(m-1)*div]
```

The example below shows how to use this function to build a planar spline curve:

```
#include "lib.h"

/* Imported function declaration */
void hermite( Flt p[], int m, Flt q0, Flt qm,
              int div, Flt output[] ) ;

/* Number of points in control polygon */
#define  M    (8)
#define DIV  (10)
/* Number of points in approximating polygon */
#define  N    ((M-1)*DIV+1)

/* Control polygon initialization */
Flt xp[M] = { 0.056, 0.287, 0.655, 0.716,
              0.228, 0.269, 0.666, 0.929 } ;
Flt yp[M] = { 0.820, 0.202, 0.202, 0.521,
              0.521, 0.820, 0.820, 0.227 } ;

Flt q0x = .1, q0y = .1 ;
Flt qmx = -.1, qmy = -.1 ;

/* Declaration of output arrays */
Flt ox[N], oy[N] ;
```

```
void main( void )
{
    /* calculate spline */
    hermite( xp, M, q0x, qmx, DIV, ox ) ;
    hermite( yp, M, q0y, qmy, DIV, oy ) ;
    /* initialize graphics mode */
    SetVideoMode() ;
    /* display control polygon */
    DrawPolygon( xp, yp, M, 1 ) ;
    /* display built approximating polygon */
    DrawPolygon( ox, oy, N, 2 ) ;
}
```

The source of the program used may be found on the applied diskette in `hermite.c` file of the subdirectory `CURVES`. The detailed information on the file names and their contents is presented in **`readme.txt`** file of directory `SPLINES` and in Appendix B.

3.5.2 Spline curves of Catmull-Rom

For given array

$$\mathbf{P}_0, \mathbf{P}_1, \mathbf{P}_2, \mathbf{P}_3$$

an (elementary) spline curve of Catmull-Rom is defined by the parametric vector equation of the form

$$\mathbf{R}(t) = \frac{1}{2}(-t(1-t)^2\mathbf{P}_0 + (2-5t^2+3t^3)\mathbf{P}_1 + t(1+4t-3t^2)\mathbf{P}_2 - t^2(1-t)\mathbf{P}_3,$$

$$0 \le t \le 1.$$

A **matrix form** of the parametric equation described the elementary spline curve of Catmull-Rom,

$$\mathbf{R}(t) = \mathbf{PMT}, \quad 0 \le t \le 1,$$

where

$$\mathbf{P} = \begin{pmatrix} \mathbf{P}_0 & \mathbf{P}_1 & \mathbf{P}_2 & \mathbf{P}_3 \end{pmatrix} = \begin{pmatrix} x_0 & x_1 & x_2 & x_3 \\ y_0 & y_1 & y_2 & y_3 \\ z_0 & z_1 & z_2 & z_3 \end{pmatrix},$$

$$\mathbf{R} = \begin{pmatrix} x(t) \\ y(t) \\ z(t) \end{pmatrix}, \quad \mathbf{M} = \frac{1}{2}\begin{pmatrix} 1 & -3 & 3 & -1 \\ 4 & 0 & -6 & 3 \\ 1 & 3 & 3 & -3 \\ 0 & 0 & 0 & 1 \end{pmatrix}, \quad \mathbf{T} = \begin{pmatrix} t^0 \\ t^1 \\ t^2 \\ t^3 \end{pmatrix}.$$

Matrix **M** is called *a basic matrix of the Catmull-Rom spline curve.*

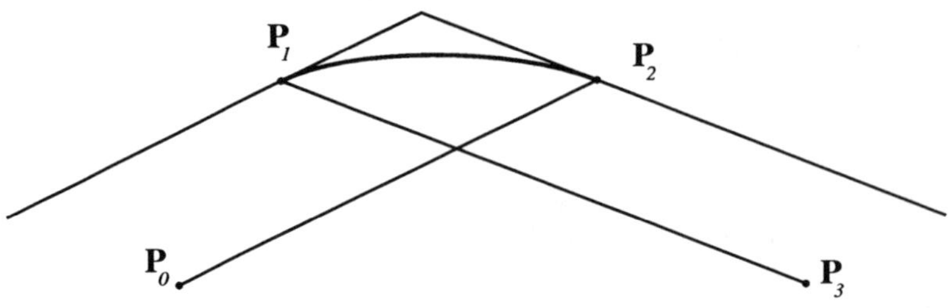

Figure 29.

A tangent of the elementary spline curve of Catmull-Rom at the end point $\mathbf{P}_1$, $\mathbf{R}(0) = \mathbf{P}_1$, is parallel to segment $\mathbf{P}_0\mathbf{P}_2$ and, at the end point $\mathbf{P}_2$, $\mathbf{R}(1) = \mathbf{P}_2$, is parallel to segment $\mathbf{P}_1\mathbf{P}_3$ (Figure 29).

Composite spline curves of Catmull-Rom. A continuous curve γ is called *a (composite) spline curve of Catmull-Rom* defined by array

$$\mathbf{P}_0, \ldots, \mathbf{P}_m, \quad m \geq 3,$$

if the curve may be represented as a union of the elementary spline curves of Catmull-Rom

$$\gamma^{(1)}, \ldots, \gamma^{(m-2)};$$

the (i)th curve $\gamma^{(i)}$ is described by the parametric equation as follows:

$$\mathbf{R}^{(i)}(t) = \begin{pmatrix} \mathbf{P}_{i-1} & \mathbf{P}_i & \mathbf{P}_{i+1} & \mathbf{P}_{i+2} \end{pmatrix} \mathbf{MT}$$

$$0 \leq t \leq 1, \quad i = 1, \ldots, m-2,$$

where $\mathbf{M}$ is the basic matrix of the Catmull-Rom spline curve.

When the composite curve γ is considered as a whole, it is more natural to use a unique parametrization for its description. Let us take an arbitrary number as a value of t_1 (for example, let $t_1 = 0$) and define the parameter knots with the formulas

$$t_{i+1} = t_i + 1, \quad i = 1, \ldots, m-3.$$

In this case an equation of the Catmull-Rom spline curve γ may be written as follows:

$$\mathbf{R} = \mathbf{R}(t), \quad t_1 \leq t \leq t_{m-2},$$

where

$$\mathbf{R} = \mathbf{R}^{(i)}(t) = \begin{pmatrix} \mathbf{P}_{i-1} & \mathbf{P}_i & \mathbf{P}_{i+1} & \mathbf{P}_{i+2} \end{pmatrix} \mathbf{M} \begin{pmatrix} (t-t_i)^0 \\ (t-t_i)^1 \\ (t-t_i)^2 \\ (t-t_i)^3 \end{pmatrix}$$

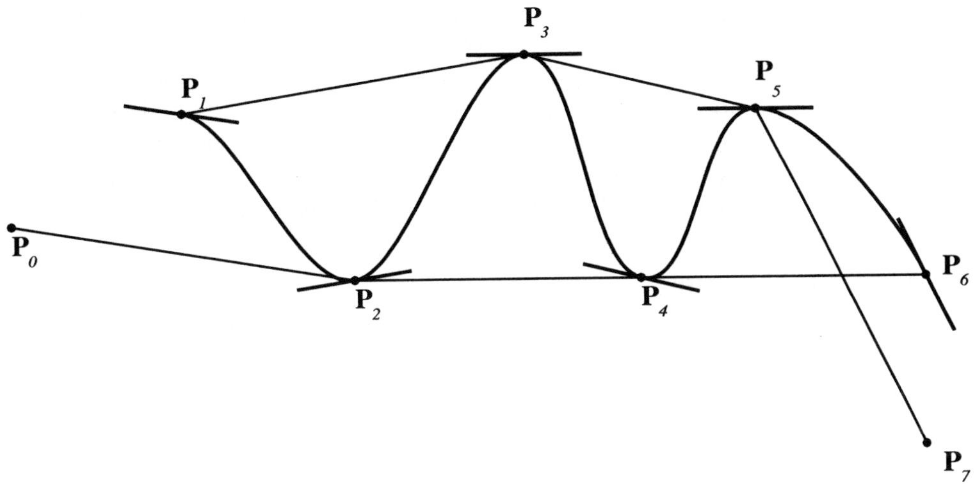

Figure 30.

$$t_i \le t \le t_{i+1}, \quad i = 1, \ldots, m-3$$

is a parametric vector equation of the (i)th elementary spline curve of Catmull-Rom $\gamma^{(i)}$.

Properties of the composite spline curve of Catmull-Rom. A composite spline curve of Catmull-Rom generated by array

$$\mathbf{P}_0, \ldots, \mathbf{P}_m, \quad m \ge 3$$

possesses the following properties:

1^+ the curve is a C^1-smooth curve.

2^+ the curve interpolates vertices $\mathbf{P}_1, \ldots, \mathbf{P}_{m-1}$; at the joining point of the elementary curves $\gamma^{(i)}$ and $\gamma^{(i+1)}$ we have

$$\mathbf{R}^{(i)}(t_{i+1}) = \mathbf{R}^{(i+1)}(t_{i+1}) = \mathbf{P}_{i-1}.$$

3^+ a tangent vector at vertex $\mathbf{P}_{i-1}$, $i = 2, \ldots, m$, is parallel to the segment which connects vertices $\mathbf{P}_{i-2}$ and $\mathbf{P}_i$,

$$\dot{\mathbf{R}}_i(t_{i+1}) = \dot{\mathbf{R}}_{i+1}(t_{i+1}) = \frac{1}{2}(-\mathbf{P}_{i-2} + \mathbf{P}_i)$$

(Figure 30).

4^- the curve does not lie in the convex hull generated by the given array.

5^+ the change of one vertex in the given array causes only part of the curve to change: when vertex $\mathbf{P}_i$ changes, it is necessary to recalculate the parametric equations of only four curves $\gamma^{(i-2)}$, $\gamma^{(i-1)}$, $\gamma^{(i)}$, $\gamma^{(i+1)}$.

6^+ with addition of one vertex to the given array, it is necessary to recalculate the parametric equations of only four elementary curves.

7^+ the composite spline curve of Catmull-Rom is affine invariant.

8^- the equations of the composite spline curve of Catmull-Rom do not contain free parameters (because of this, the curve is uniquely determined by the given array of vertices and there is no way of controlling its shape).

9^- the spline curve of Catmull-Rom is projectively noninvariant.

Implementation

The following C function is an implementation of the algorithm described above.

Function syntax:

```
catmull_rom( p, m, div, output )
```

Function parameters are:

```
Flt p[]        (in)  - control polygon of spline
                       elements p[0] .. p[ m-1 ]
int m          (in)  - number of points in control polygon
int div        (in)  - number of links in approximating
                       polygon
Flt output[] (out) - approximating polygon
                       elements output[0] .. output[(m-3)*div]
```

The example below shows how to use this function to build a planar spline curve:

```
#include "lib.h"

/* Imported function declaration */
void catmull_rom( Flt p[], int m, int div, Flt output[] ) ;

/* number of points in control polygon */
#define  M    (8)
#define DIV  (10)
/* number of points in approximating polygon */
#define  N    ((M-3)*DIV)

/* Control polygon initialization */
Flt xp[M] = { 0.056, 0.287, 0.655, 0.716,
              0.228, 0.269, 0.666, 0.929 } ;
Flt yp[M] = { 0.820, 0.202, 0.202, 0.521,
              0.521, 0.820, 0.820, 0.227 } ;

/* Declaration of output arrays */
```

```
Flt ox[(M-3)*DIV], oy[(M-3)*DIV] ;

void main( void )
{
    /* calculate spline */
    catmull_rom( xp, M, DIV, ox ) ;
    catmull_rom( yp, M, DIV, oy ) ;
    /* initialize graphics mode */
    SetVideoMode() ;
    /* display control polygon */
    DrawPolygon( xp, yp, M, 1 ) ;
    /* display built approximating polygon */
    DrawPolygon( ox, oy, N, 2 ) ;
}
```

The source of the program used may be found on the applied diskette in `catmull.c` file of the subdirectory `CURVES`. The detailed information on the file names and their contents is presented in `readme.txt` file of directory `SPLINES` and in Appendix B.

3.5.3 Composite planar implicit cubic curves

The planar spline curves may be generated by the elementary fragments of the algebraic curves defined by the implicit equations.

Geometric construction of the fragment of the algebraic curve of the third order. Let $\mathbf{P}_0$, $\mathbf{P}_1$, $\mathbf{P}_2$ be noncollinear points. The disposition of the points on the plane with respect to the triangle $\Delta = \Delta\mathbf{P}_0\mathbf{P}_1\mathbf{P}_2$ may be described with the help of *barycentric coordinates.*

Let (x_i, y_i) be Cartesian coordinates of the points $\mathbf{P}_i$, $i = 0, 1, 2$, and $\mathbf{P}(x, y)$ an arbitrary point on the plane. Then the following decomposition

$$(x, y) = s(x_0, y_0) + t(x_1, y_1) + u(x_2, y_2)$$

is valid (here s, t and u, $s + t + u = 1$, are the *barycentric coordinates* of the point $\mathbf{P}(x, y)$ with respect to the points $\mathbf{P}_0$, $\mathbf{P}_1$ and $\mathbf{P}_2$). If $\mathbf{P}(x, y)$ is an inner point of the triangle Δ, then $s > 0$, $t > 0$, $u > 0$.

Let $\mathbf{A}(s_0, t_0, u_0)$ be an arbitrary inner point of the triangle Δ ($s_0 > 0$, $t_0 > 0$, $u_0 > 0$) and point $\mathbf{B}(s_1, 0, u_1)$ lie on the extension of the segment $\mathbf{P}_0\mathbf{P}_2$ ($s_1 < 0$, $u_1 > 1$ or $s_1 > 1$, $u_i < 0$) (Figure 31).

In the family of the implicit algebraic curves of the third degree

$$as^2 + bsu^2 + cst^2 + dt^2u + estu = 0$$

there exists a curve which passes through point $\mathbf{P}_0$ being tangent to segment $\mathbf{P}_0\mathbf{P}_1$, through point $\mathbf{P}_2$ being tangent to segment $\mathbf{P}_1\mathbf{P}_2$, and through

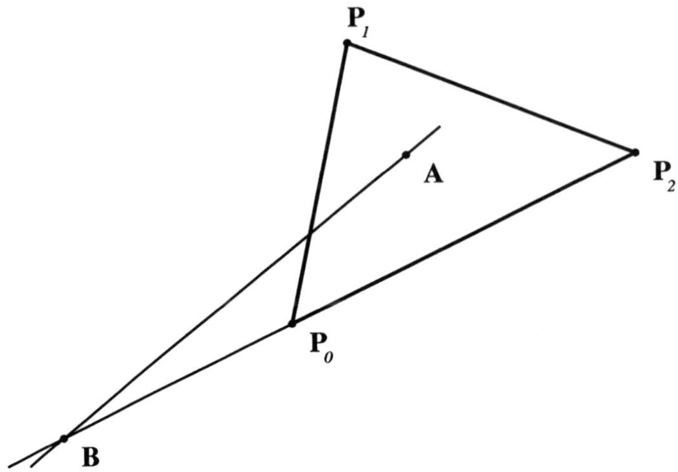

Figure 31.

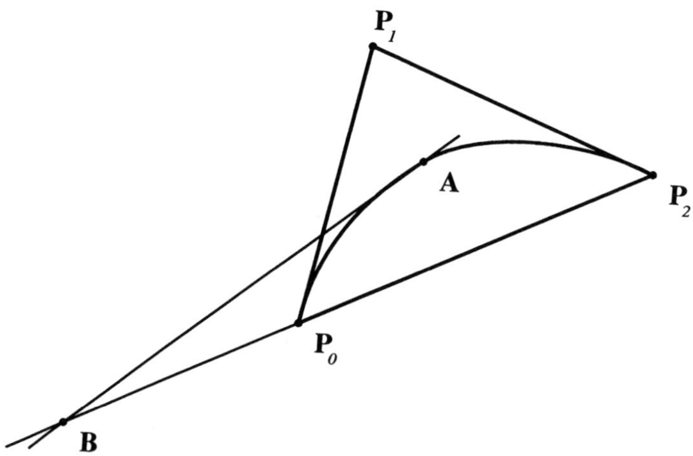

Figure 32.

point **A** being tangent to segment **AB** (Figure 32). To write an equation of the curve, it is sufficient to assume

$$a = -h\beta_2 t_0^2 u_0, \quad b = \beta_1 s_0 t_0^2, \quad c = -(1-\beta_1) s_0 u_0^2,$$
$$d = h(1-\beta_2) s_0^2 u_0, \quad e = (1 - h - 2\beta_1 + 2h\beta_2) s_0 t_0 u_0,$$

where $0 < \beta_1, \beta_2 < 1$ are arbitrary numbers and

$$h = \frac{s_0 u_1}{s_1 u_0}.$$

A curvature of the curve segment lying in the triangle is nonnegative.

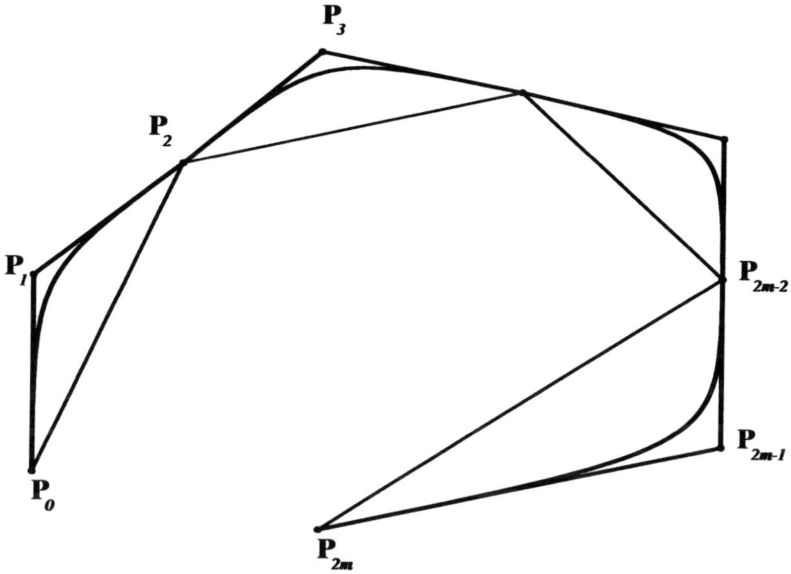

Figure 33.

Composite algebraic curve. Let

$$\mathbf{P}_0, \mathbf{P}_1, \ldots, \mathbf{P}_{2m},$$

be a planar array of which every three points $\mathbf{P}_{2i-1}$, $\mathbf{P}_{2i}$, $\mathbf{P}_{2i+1}$, $i = 1, \ldots, m-1$, are collinear (Figure 33).

Choose inside each triangle

$$\Delta_i = \Delta \mathbf{P}_{2i} \mathbf{P}_{2i+1} \mathbf{P}_{2i+2}$$

an arbitrary interpolated point $\mathbf{A}_i$, a tangent to the desired curve at the point and the shape parameters $\beta_1^{(i)}$ and $\beta_2^{(i)}$, and consider the curve γ composed of the elementary fragments $\gamma^{(i)}$ constructed for every triangle Δ_i according to the rule described above. The composite curve will be G^1-continuous and convex in every triangle Δ_i.

The choice of the points, the tangents, and parameters may be universal for all of the triangles Δ_i. For example, one can take $\beta_1 = \beta_2 = 1/2$, $h = -1/3$, and the gravity centers of the triangles Δ_i as the points $\mathbf{A}_i$.

Chapter 4

Spline surfaces

The general problem of this chapter may be formulated as follows: construct a smooth surface for a given array of vertices accounting for their enumeration so that the surface running quietly passes near these vertices and is subject to some additional conditions. The conditions may be of various character. For example, one can require the desired surface to pass through all of the given vertices, to be tangent to given directions, or to be of given regularity, etc.

In the search for the appropriate solution of the problem, a polyhedral surface, the vertices of which coincide with the points of the given array, plays an important role. The polyhedral surface is called *a control* or *support polyhedral surface*. In many cases the polyhedral surface discloses the course of the unknown surface sufficiently well. The property is of prime importance in solving the smoothing problem.

We shall consider the arrays organized as a two-dimensional network topologically equivalent to the tamely imbedded rectangular net. This implies that in array

$$\begin{matrix} \mathbf{P}_{00}, & \mathbf{P}_{01}, & \ldots, & \mathbf{P}_{0n}, \\ \ldots, & \ldots, & \ldots, & \ldots, \\ \mathbf{P}_{m0}, & \mathbf{P}_{m1}, & \ldots, & \mathbf{P}_{mn} \end{matrix}$$

hand in hand with the vertices $\mathbf{P}_{ij}$ the edges connecting them

$$\mathbf{P}_{ij}\mathbf{P}_{i,j+1}, \quad i = 0, 1, \ldots, m-1, m; \quad j = 0, 1, \ldots, n-1,$$

$$\mathbf{P}_{ij}\mathbf{P}_{i+1,j}, \quad i = 0, 1, \ldots, m-1; \quad j = 0, 1, \ldots, n-1, n$$

are taken into account (Figure 1). The vertices

$$\mathbf{P}_{ij}, \quad i = 1, \ldots, m-1; \quad j = 1, \ldots, n-1$$

are called *inner vertices* and the vertices

$$\begin{matrix} \mathbf{P}_{0j} & (j = 0, 1, \ldots, n-1), & \mathbf{P}_{mj} & (j = 1, 2, \ldots, n), \\ \mathbf{P}_{i0} & (i = 1, 2, \ldots, m), & \mathbf{P}_{in} & (i = 0, 1, \ldots, m-1) \end{matrix}$$

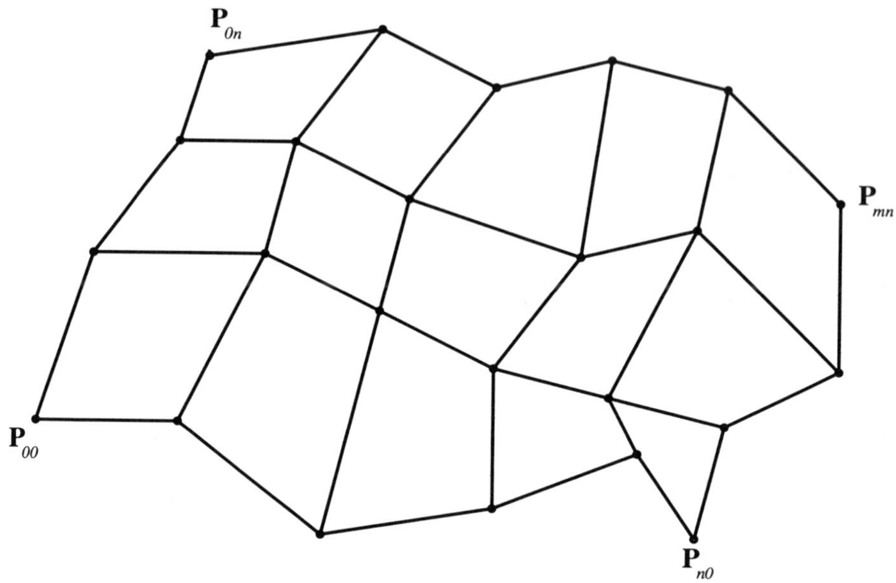

Figure 1.

are called *boundary vertices.* Every inner vertex has four neighboring vertices, and every boundary vertex has three of them except for a quadruple of *corner vertices*

$$\mathbf{P}_{00}, \quad \mathbf{P}_{0n}, \quad \mathbf{P}_{mn}, \quad \mathbf{P}_{m0}$$

(each has only two neighboring vertices).

The array organized in such a way is often called *a control* or *support gragh* of the desired spline surface.

Contrary to the case considered in the second chapter, here, the restrictions on the set of vertices are less stringent. Because of this, the description of a desired surface should be searched in a more general parametric form.

To describe the appropriate parametric equations of the desired surface, it is convenient to use the way connected naturally with the intuitive concept of the surface as a trace of the varying curve which moves along the other curve and changes its shape. In addition, it is desirable to take into account a structure of the control gragh.

Let us attempt to elucidate a geometric meaning of the parametric equation of the form

$$\mathbf{R}(u,v) = \sum_{i=0}^{m} \sum_{j=0}^{n} a_i(u) b_j(v) \mathbf{P}_{ij}, \tag{$*$}$$

where $a_i(u)$ and $b_j(v)$ are certain functional coefficients.

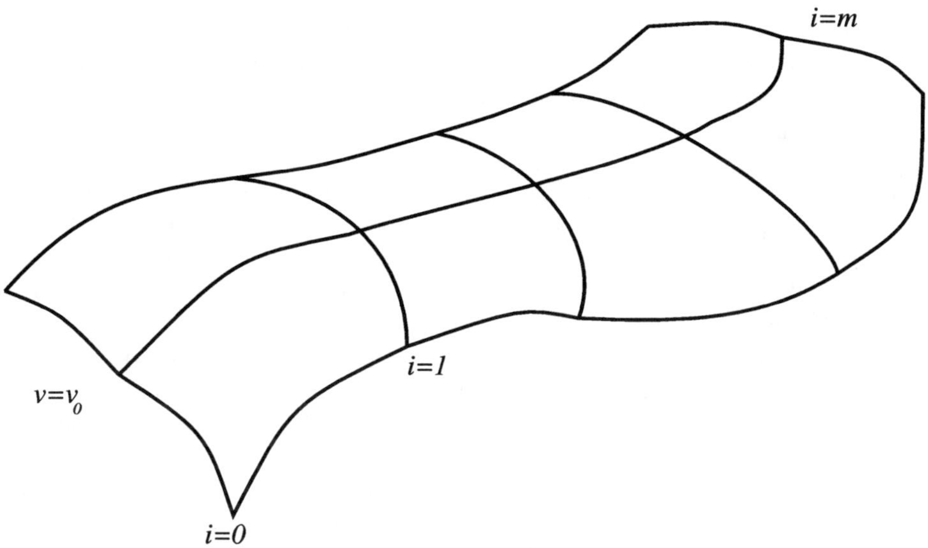

Figure 2.

Let

$$\mathbf{R}_i(v) = \sum_{j=0}^{n} b_j(v)\mathbf{P}_{ij}, \quad i = 0, \ldots, m.$$

Then the equation $(*)$ may be rewritten as follows

$$\mathbf{R}(u, v) = \sum_{i=0}^{m} a_i(u)\mathbf{R}_i(v).$$

The first of the relations may be considered as a parametric equation of the ith curve ($i = 0, \ldots, m$). Fix the variable v, $v = v_0 = \text{const}$. As a consequence we have a marked point on each of the curves (point $\mathbf{R}_i(v_0)$ on ith curve). For the received point set, define a curve by the parametric equation

$$\mathbf{R}(u, v_0) = \sum_{i=0}^{m} a_i(u)\mathbf{R}_i(v_0)$$

(Figure 2).

The equation $(*)$ is said to be *a tensor product description* of the surface. Such *tensor-product surfaces* possess a number of advantages. One of them is most essential for our purposes because it makes it possible to extend many results and observations from the previous chapter devoted to curves to the two-dimensional case. The fact is that a great number of new questions arise while the dimension of the problem increases. The restrictions on the structure of the support arrays and the choice of the most simple classes of surfaces, as the working classes, permit one to keep the number within the bounds deemed reasonable for the first acquaintance.

If the vertex number in the given array **P** is large enough, searching for the universal functional coefficients $c_{ij}(u,v) = a_i(u)b_j(v)$, as a rule, may be rather difficult. In addition, it may turn out (every so often such is indeed the case) that the found functions $c_{ij}(u,v)$, in parallel with the desired properties, have properties which are not easily compatible with the expected behavior of the corresponding surface (for example, the surface described by equation $(*)$ with such coefficients may oscillate or deviate greatly from the given set).

To solve the formulated problem, we can conveniently use the curves composed of the elementary fragments. If the elementary fragments are constructed according to a unified relatively simple scheme, the corresponding composite surfaces are called *spline surfaces.*

The parametric equations of every elementary fragment are sought in the form $(*)$, and the functional coefficients are of the same nature (for example, polynomials of the same degree), but for constructing a special fragment of the desired surface, we use only part of the given vertices from the set **P**.

The functional coefficients $c_{ij}(u,v)$, where the factors $a_i(u)$ and $b_j(v)$ are the polynomials of the small (second or third) orders, are usually used for the description of the elementary surfaces and the calculation of their geometric characteristics (the information is essential for joining). The reason for this is that the coefficients are calculated relatively easily. Of course, with the help of the polynomials of the higher orders, one can describe the complex-shaped surface, but such polynomials have many coefficients whose geometric and physical meaning is very difficult to understand. In addition, using the polynomials of the higher degrees may initiate undesired oscillations of the resulting surface.

The most widespread ways of the composite surface construction are the methods involving the use of bicubic polynomials (by the way, the polynomials were extensively used in the second chapter). The choice allows one to account for both the differential and external-geometric requirements imposed on the desired surface.

The above defines the way the chapter sections are presented. The chapter begins with the description of the minimal necessary information from the differential geometry of surfaces. Then we consider various surface classes which will help resolve the above formulated approximation problem: the elementary surfaces (the future fragments) are defined and their main properties are described. The approach will help a user in the choice of the most appropriate class of surfaces. Part of the properties is formulated in the form of answers to a standard set of questions, including such questions as:

What is the smoothness order of the constructed surface?

What are the distinguishing features of the surface arrangement with respect to the given vertices of the control graph?

How does the change of the support vertices influence the surface shape?

It is useful to know how the surface shape and the simplest geometric transformations (affine and perspective) are related to each other.

Recall that the affine transformations include a rotation, a dilation, a translation and their combinations, and the projective transformations include a transformation of the perspective as well.

Most of the classes under consideration possess the affine invariance property. Among them exist the projective invariants. While the final result is independent of the fulfillment sequence of these two operations (the transformation of the support array and the construction of the surface), nevertheless the choice of the operation's sequence may substantially reduce the volume of calculations.

Use of the rational surfaces introduces the wanted adaptability into solution of the approximation problem because the *free* numerical parameters take part in description of the surfaces. The possibility of a choice permits influence on the shape of the desired surfaces and account of the prescribed properties. An important point is that the rational surfaces hold many of the properties of the corresponding nonrational (polynomial) surfaces. The projective invariance of the rational surfaces is their most attractive property (also very useful in visualization).

Besides, we show how one can achieve a composite surface with the desired properties by joining the elementary surfaces and discuss various questions connected with the program implementation of the considered approaches (characteristic properties, advantages and disadvantages, stability questions, etc).

4.1 Main facts from the surface theory

4.1.1 Parametrized surfaces

A set S of points $M(x, y, z)$ whose Cartesian coordinates x, y, and z are defined by formulas

$$x = x(u, v), \quad y = y(u, v), \quad z = z(u, v),$$

$$a \leq v \leq b, \quad c \leq v \leq d,$$

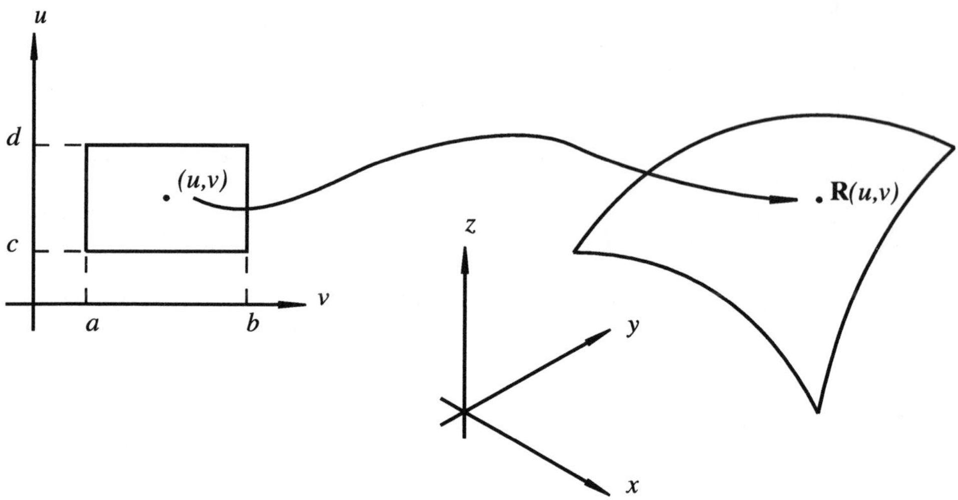

Figure 3.

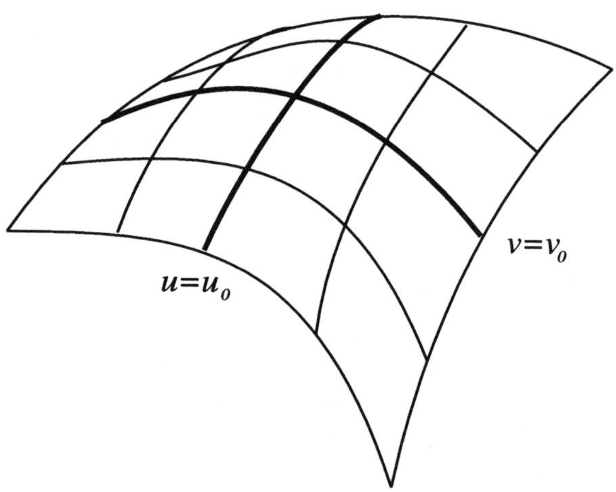

Figure 4.

where functions $x(u,v)$, $y(u,v)$, and $z(u,v)$ are continuous in rectangle

$$R = \{(u,v) \mid a \leq y \leq b, \quad c \leq v \leq d\}$$

or in *vector-matrix form*,

$$\mathbf{R} = \mathbf{R}(u,v) = \begin{pmatrix} x(u,v) \\ y(u,v) \\ z(u,v) \end{pmatrix},$$

$$(u, v) \in R$$

is called *a parametrically defined (parametrized) surface.*

The relations are called *parametric equations of surface S* or simply *a parametrization of surface S* (Figure 3).

Variables u and v are called *inner curvilinear coordinates on surface S.* When $v = v_0 = \text{const}$ vector equation

$$\mathbf{R} = \mathbf{R}(u, v_0), \quad c \leq v \leq d$$

describes a curve on surface S — *u-line.* Analogously, when $u = u_0 = \text{const}$ we have vector equation

$$\mathbf{R} = \mathbf{R}(u_0, v), \quad a \leq u \leq b$$

of *v-line*; u-lines and v-lines construct *a coordinate net on surface S* (Figure 4).

4.1.2 Smooth and regular surfaces

We say that surface S is *C^r-smooth with respect to the given parametrization* if vector-function $\mathbf{R}(u, v)$ is C^r-smooth in rectangle R, that is, each of coordinate functions $x(u, v)$, $y(u, v)$, and $z(u, v)$ has continuous derivatives up to and including r in the rectangle.

A smooth parametrization is named *regular* if

$$\mathbf{R}_u(u, v) \wedge \mathbf{R}_v(u, v) \neq \mathbf{O}, \quad (u, v) \in R.$$

For regular surface S vectors $\mathbf{R}_u(u, v)$ and $\mathbf{R}_v(u, v)$ are tangent to u-line and v-line at point $\mathbf{R}(u, v)$, respectively, lie in its *tangent plane* at the point, and a unit vector

$$\mathbf{N} = \frac{\mathbf{R}_u \wedge \mathbf{R}_v}{| \mathbf{R}_u \wedge \mathbf{R}_v |}$$

is parallel to *a normal* to surface S at the point.

A triple of noncoplanar vectors

$$\mathbf{R}_u, \quad \mathbf{R}_v, \quad \mathbf{N}$$

are defined at every point of the regular surface (Figure 5).

4.1.3 First fundamental form of a surface

A squared arc element

$$\mathrm{I} = d\mathbf{R}^2 = (\mathbf{R}_u du + \mathbf{R}_v dv)^2 = E du^2 + 2F du dv + G dv^2,$$

where $E = E(u, v) = \mathbf{R}_u^2$, $F = F(u, v) = \mathbf{R}_u \mathbf{R}_v$, $G = G(u, v) = \mathbf{R}_v^2$ is called *the first fundamental form* of regular surface S and functions E, F, and G are *coefficients of the first fundamental form.* The coefficients are connected by inequality

$$EG - F^2 > 0.$$

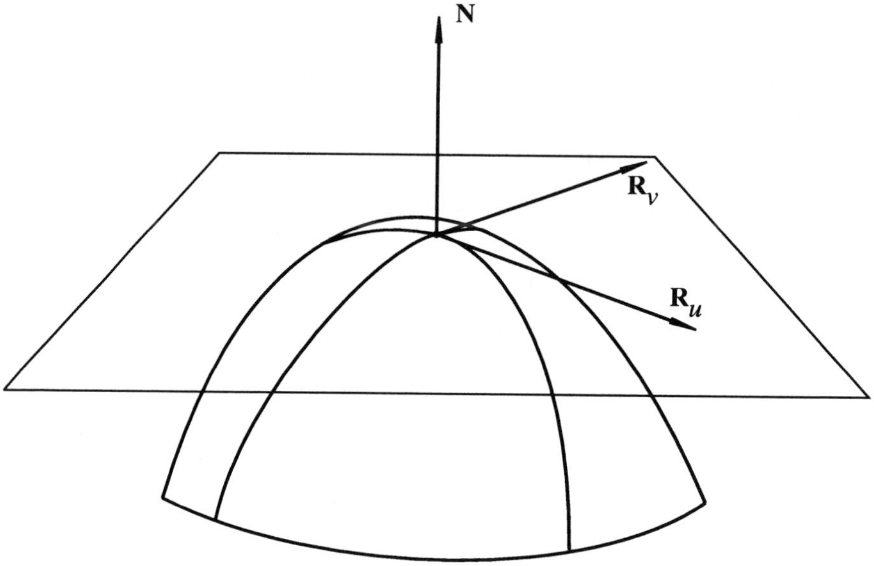

Figure 5.

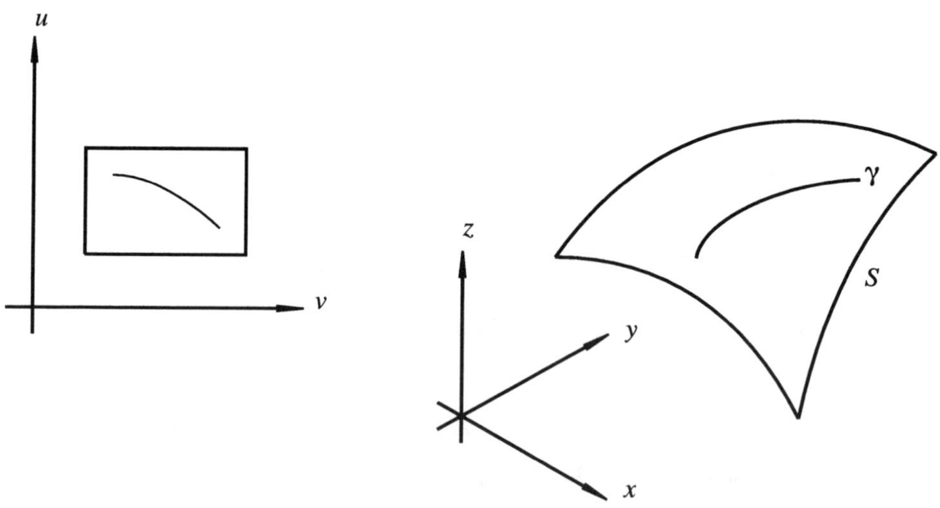

Figure 6.

4.1.4 Curve on a surface

The parametric equations

$$u = u(t), \quad v = v(t), \quad \alpha \le t \le \beta$$

define a curve γ on the surface S. The curve is described by vector equation

$$\gamma : \mathbf{R} = \mathbf{R}(t) = \mathbf{R}(u(t), v(t)), \quad \alpha \le t \le \beta$$

(Figure 6). The equations

$$u = u(t), \quad v = v(t)$$

are called *inner equations of the curve.*

The curve γ is regular on the regular surface S when

$$\dot{u}^2 + \dot{v}^2 > 0.$$

A tangent vector of curve γ may be calculated by formula

$$\dot{\mathbf{R}} = \mathbf{R}_u \dot{u} + \mathbf{R}_v \dot{v}$$

and an arc length by formula

$$s = s(\xi) = \int_\alpha^\xi \mid \dot{\mathbf{R}}(t) \mid dt = \int_\alpha^\xi \sqrt{E\dot{u}^2 + 2F\dot{u}\dot{v} + G\dot{v}^2} dt.$$

4.1.5 Angle between curves on a surface

Let curves described by parametric equations

$$u = u_1(t), \quad v = v_1(t) \quad \text{and} \quad u = u_2(t), \quad v = v_2(t)$$

intersect at point $M(t)$ of the parametric plane corresponding to the parameter value t (Figure 7).

The corresponding curves on surface S

$$\gamma_1 : \mathbf{R} = \mathbf{R}(u_1(t), v_1(t)) \quad \text{and} \quad \gamma_2 : \mathbf{R} = \mathbf{R}(u_2(t), v_1(t))$$

intersect at the right angle at point $\mathbf{R}(t)$ of surface S if

$$E\dot{u}_1\dot{u}_2 + F(\dot{u}_1\dot{v}_2 + \dot{u}_2\dot{v}_1) + G\dot{v}_1\dot{v}_2 = 0.$$

Coordinate u-lines and v-lines on surface S are orthogonal if and only if $F(u, v) = 0$.

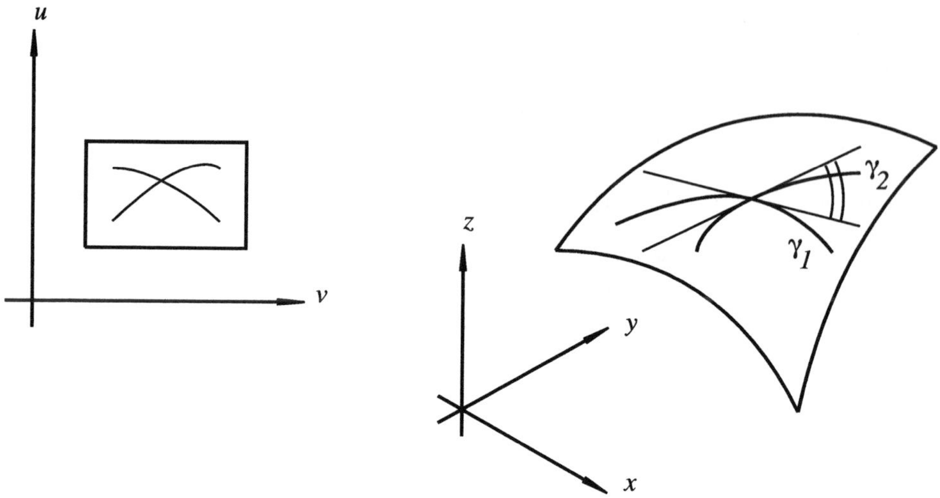

Figure 7.

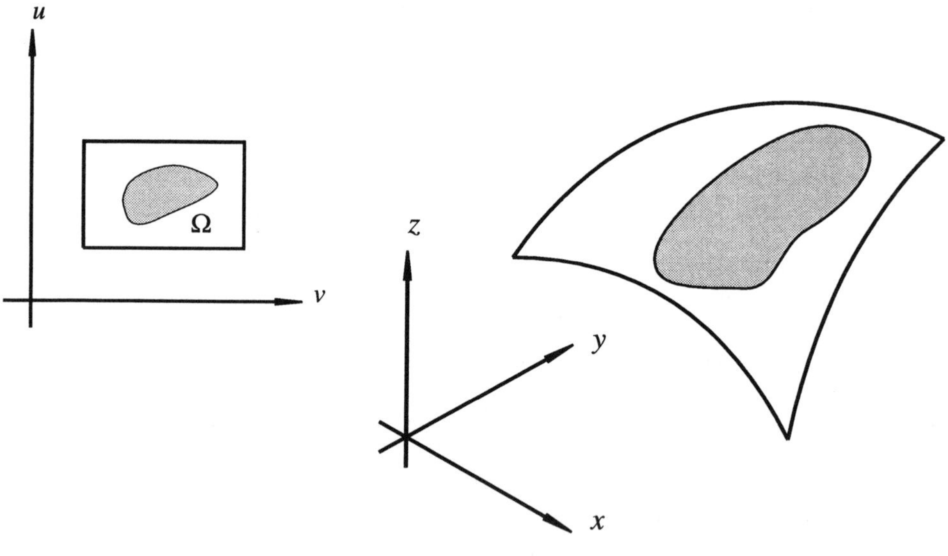

Figure 8.

4.1.6 Area of a surface

An area of a patch on surface S corresponding to domain Ω of the parametric plane (u, v) (Figure 8) is equal to

$$S = \int\int_{\Omega} \sqrt{EG - F^2}dudv.$$

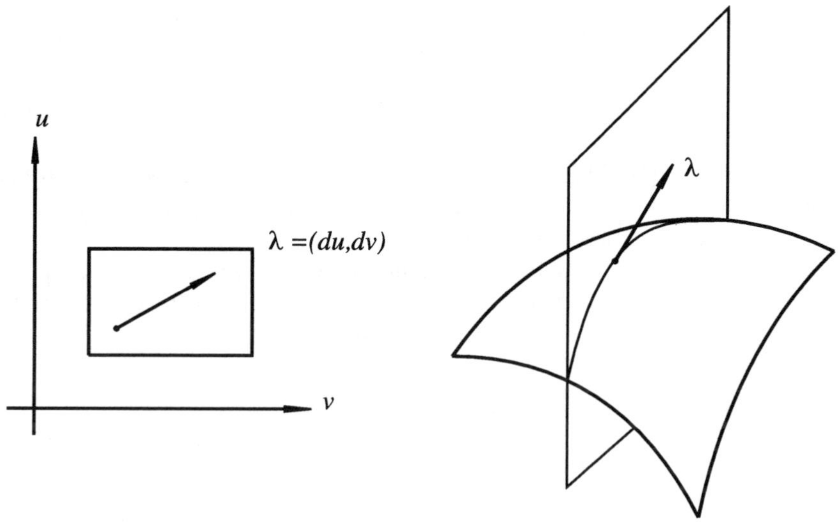

Figure 9.

4.1.7 Second fundamental form of a surface

An expression

$$\mathrm{II} = d^2\mathbf{R}\mathbf{N} = Ldu^2 + 2Mdudv + Ndv^2,$$

where

$$L = \frac{(\mathbf{R}_{uu} \wedge \mathbf{R}_u \wedge \mathbf{R}_v)}{\sqrt{EG - F^2}} = (\mathbf{R}_{uu}\mathbf{N}),$$

$$M = \frac{(\mathbf{R}_{uv} \wedge \mathbf{R}_u \wedge \mathbf{R}_v)}{\sqrt{EG - F^2}} = (\mathbf{R}_{uv}\mathbf{N}),$$

$$N = \frac{(\mathbf{R}_{vv} \wedge \mathbf{R}_u \wedge \mathbf{R}_v)}{\sqrt{EG - F^2}} = (\mathbf{R}_{vv}\mathbf{N})$$

is called *the second fundamental form* of surface S.

4.1.8 Lines of curvature

A curvature of a curve lying on surface S and passing through point $\mathbf{R}(u, v)$ in direction $\lambda = (du : dv)$ may be calculated by formula

$$k(\lambda) = k(\mathbf{R}, \lambda) = \frac{Ldu^2 + 2Mdudv + Ndv^2}{Edu^2 + 2Fdudv + Gdv^2}$$

(Figure 9).

Two following cases are possible here.

1. Function $k(\mathbf{R}, \lambda)$ is independent of the direction λ at point $\mathbf{R}$. Then the point is called either *an umbilical point* or *a planar point* of the surface S.

All the points of a sphere are umbilic; all the points of a plane are planar.

2. Function $k(\mathbf{R}, \lambda)$ has just two extreme values k_- and k_+ — *principal curvatures* of surface S at the given point. The corresponding directions on the surface are called *principal directions.*

The net of curves that have the principal directions at all of the points is called the net of *lines of curvature.* If $F = 0$ and $M = 0$, then the net of the coordinate lines is the net of the curvature lines.

4.1.9 Gaussian and mean curvatures

A product of the principal curvatures at the given point of the surface is called its *Gaussian curvature* at the point,

$$K = k_- k_+ = \frac{LN - M^2}{EG - F^2}$$

and their halfsum is called its *mean curvature,*

$$H = \frac{1}{2} = \frac{NE - 2MF + LG}{EG - F^2}.$$

The Gaussian curvature and the mean curvature calculated at a given point of the regular surface describe a shape and a measure of its deviation from a tangent plane at the point. A point of the surface is called *elliptic* if $K > 0$ at the point, *hyperbolic* if $K < 0$, and *parabolic* if $K = 0$.

4.1.10 Geometric continuity

It is important to keep in mind the following two circumstances. On the one hand, the surface as a set of points in the space may possess excellent geometric characteristics but be given by the relatively poor parametric equations, and on the other, the surface may have the coordinate functions with high differential properties and be a nonregular surface.

Let us consider some examples.

Example 1. Surface S is given by parametric equation

$$\mathbf{R} = \mathbf{R}(u, v) = \begin{pmatrix} u^5 \cos v \\ u^5 \sin v \\ u^5 \end{pmatrix}, \quad -1 \le u \le 1, \quad 0 \le v \le 2\pi$$

(Figure 10).

Result: the coordinate functions are infinitely differentiable and the surface has a singularity — *a conic point.*

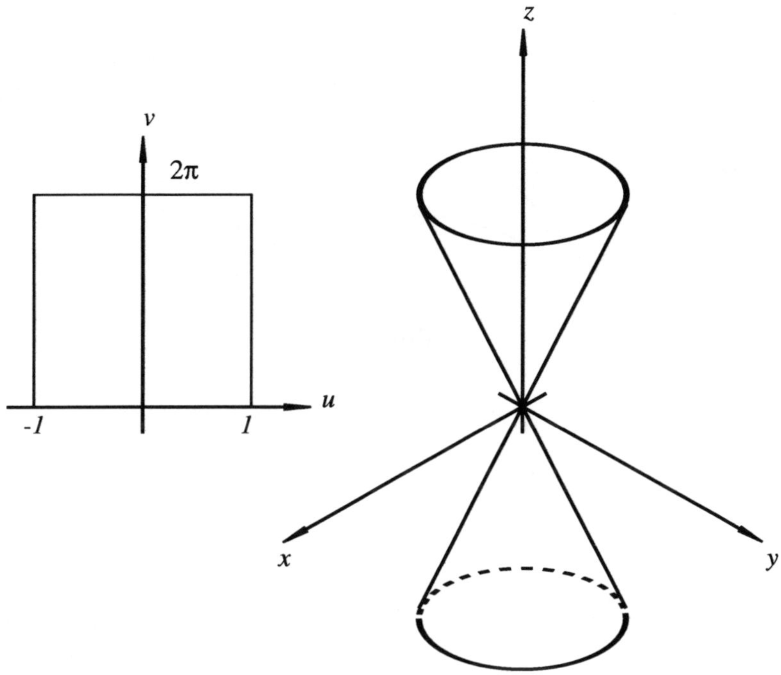

Figure 10.

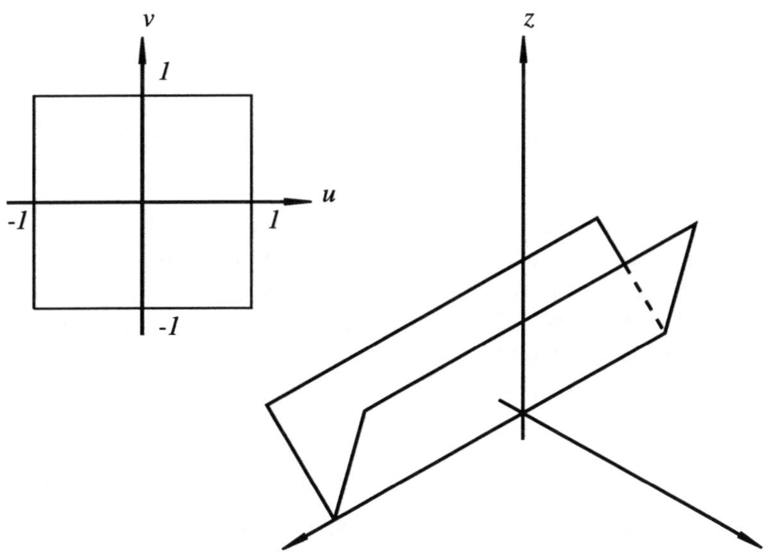

Figure 11.

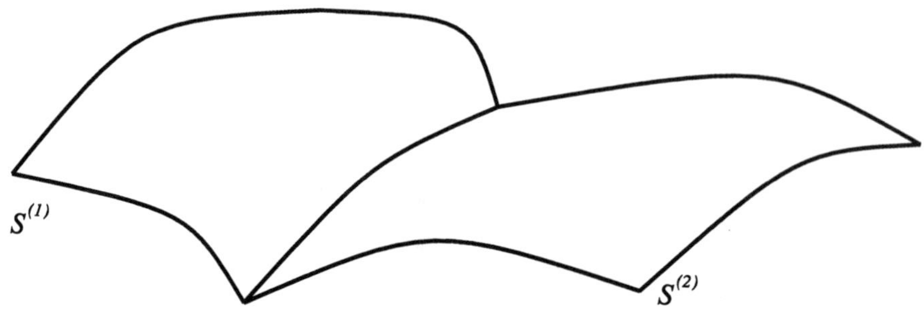

Figure 12.

Example 2. Surface S is given by the parametric equation

$$\mathbf{R} = \mathbf{R}(u, v) = \begin{pmatrix} v \\ u^5 \\ u^4|u| \end{pmatrix}, \quad -1 \le u \le 1, \quad -1 \le v \le 1$$

(Figure 11).

Result: the coordinate functions have continuous derivatives up to and including the fourth order and the surface has a singularity — *an edge.*

Cause of singularities in both cases is the same:

$$\mathbf{R}_u \wedge \mathbf{R}_v = \mathbf{O}.$$

The requirement of regularity,

$$\mathbf{R}_u \wedge \mathbf{R}_v \neq \mathbf{O},$$

is a sufficient condition for the absence of the edges and the conic points on the surface.

Every so often one can come up against a situation where each regular surface taking part in construction of a new composite surface has its own parametrization. The parametrization nonconformity may be (and often is) a cause of singularities along the joining line. The singularities are the *singularities of parametrization.*

Consider two regular surfaces $S^{(1)}$ and $S^{(2)}$ described by the parametric equations:

$$\mathbf{R} = \mathbf{R}^{(1)}(u, v), \quad a^{(1)} \le u \le b^{(1)}, \quad c^{(1)} \le v \le d^{(1)},$$

$$\mathbf{R} = \mathbf{R}^{(2)}(u, v), \quad a^{(2)} \le u \le b^{(2)}, \quad c^{(2)} \le v \le d^{(2)},$$

respectively, and such that their boundary curves have a common segment (Figure 12). In this case a composite surface $S = S^{(1)} \cup S^{(2)}$ is called *continuous*, or *G^0-continuous.*

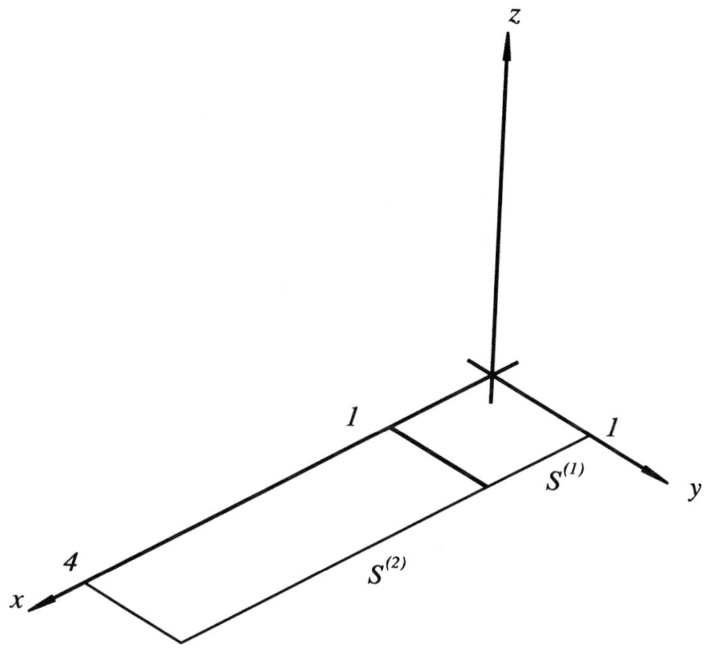

Figure 13.

A G^0-continuous composite surface

$$S = S^{(1)} \cup S^{(2)}$$

is called G^1*-continuous*, or G^1*-surface* if its tangent plane varies continuously along the common boundary curve.

Any G^1-surface may be parametrized so that it will be C^1-regular with respect to the parametrization.

Example 3. A composite surface defined by equations

$$\mathbf{R} = \mathbf{R}^{(1)}(u,v) = \begin{pmatrix} u \\ v \\ 0 \end{pmatrix}, \quad 0 \le u \le 1, \quad 0 \le v \le 1,$$

$$\mathbf{R} = \mathbf{R}^{(2)}(u,v) = \begin{pmatrix} (1+u)^2 \\ v \\ 0 \end{pmatrix}, \quad 0 \le u \le 1, \quad 0 \le v \le 1$$

is a G^1-surface. It is simply a patch of the coordinate plane $0xy$ (Figure 13).

A G^1-continuous composite surface $S = S^{(1)} \cup S^{(2)}$ is called G^2*-continuous*, or a G^2*-surface* if its Gaussian curvature varies continuously along the common boundary curve.

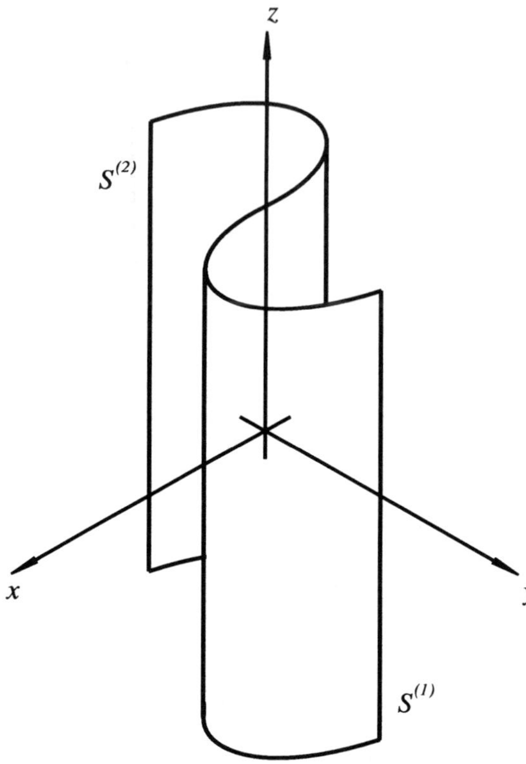

Figure 14.

Example 4. A composite surface defined by equations

$$\mathbf{R} = \mathbf{R}^{(1)}(u, v) = \begin{pmatrix} u \\ u^3 \\ v \end{pmatrix}, \quad 0 \le u \le 1, \quad 0 \le v \le 1,$$

$$\mathbf{R} = \mathbf{R}^{(2)}(u, v) = \begin{pmatrix} -\frac{2}{\pi}\cos\left(\frac{\pi}{2}u\right) \\ -\left(\frac{2}{\pi}\right)^3\cos^3\left(\frac{\pi}{2}u\right) \\ \frac{1}{2}v^2 \end{pmatrix}, \quad 0 \le u \le 1, \quad 0 \le v \le 1$$

is a G^2-surface (Figure 14).

For any G^2-surface there exists parametrization with respect to which the surface will be C^2-regular.

4.1.11 Twist vector and bilinear surface

Let a regular surface S be given by the parametric vector equation

$$\mathbf{R} = \mathbf{R}(u, v).$$

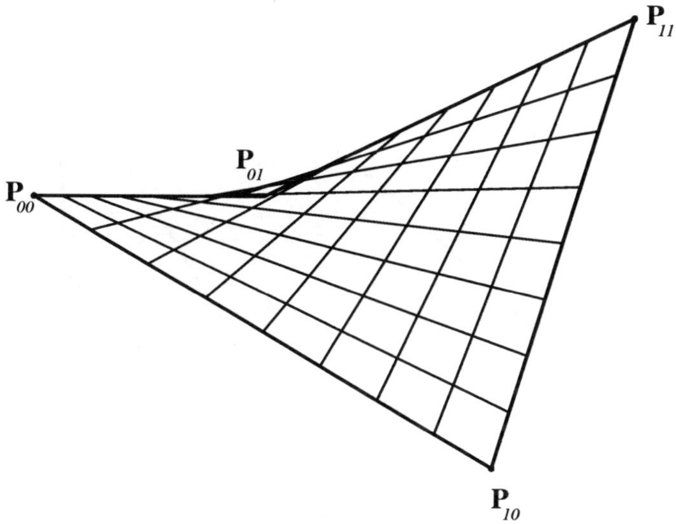

Figure 15.

A mixed derivative $\mathbf{R}_{uv}$ is called *a twist vector* of the surface.

It is best to unveil the *geometric meaning* of the twist vector using a bilinear surface as an example.

A surface described by the parametric equations

$$\mathbf{R}(u,v) = \begin{pmatrix} 1-u & u \end{pmatrix} \begin{pmatrix} \mathbf{P}_{00} & \mathbf{P}_{01} \\ \mathbf{P}_{10} & \mathbf{P}_{11} \end{pmatrix} \begin{pmatrix} 1-v \\ v \end{pmatrix},$$

$$0 \le u, v \le 1,$$

where

$$\mathbf{P}_{00}, \quad \mathbf{P}_{01}, \quad \mathbf{P}_{10}, \quad \mathbf{P}_{11}$$

is a quadruple of the noncoplanar vertices, is called *a bilinear surface* $\mathcal{B}$, defined by the quadruple (Figure 15).

The boundary curves of the surface $\mathcal{B}$ are the straight-line segments

$$\mathbf{P}_{00}\mathbf{P}_{10}, \quad \mathbf{P}_{10}\mathbf{P}_{11}, \quad \mathbf{P}_{11}\mathbf{P}_{01}, \quad \mathbf{P}_{01}\mathbf{P}_{00}.$$

The twist vector of the bilinear surface $\mathcal{B}$ may be readily calculated

$$\mathbf{R}_{uv}(u,v) = \mathbf{P}_{00} - \mathbf{P}_{01} + \mathbf{P}_{11} - \mathbf{P}_{10}.$$

The vector shows the deviation measure of point $\mathbf{P}_{11}$ from a tangent plane of the surface $\mathcal{B}$ at point $\mathbf{P}_{00}$ (Figure 16).

Important remark. A value of the mixed derivative of the radius-vector depends not only on the geometric properties of the surface but largely on the way of its parametrization.

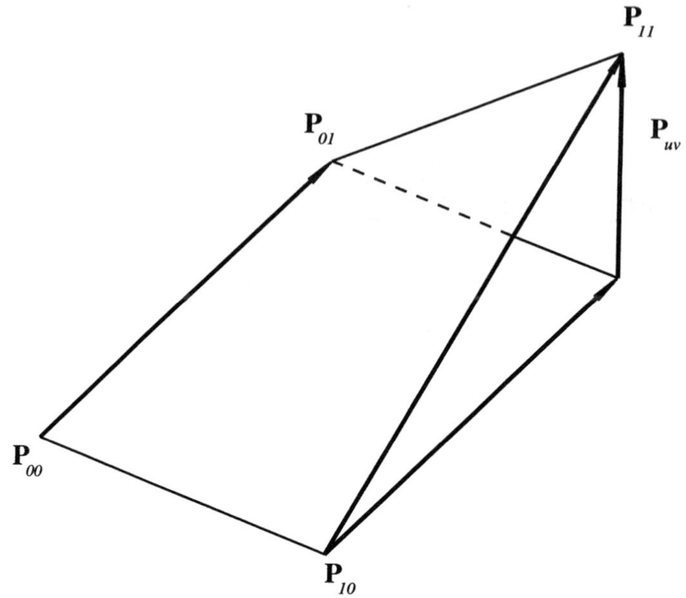

Figure 16.

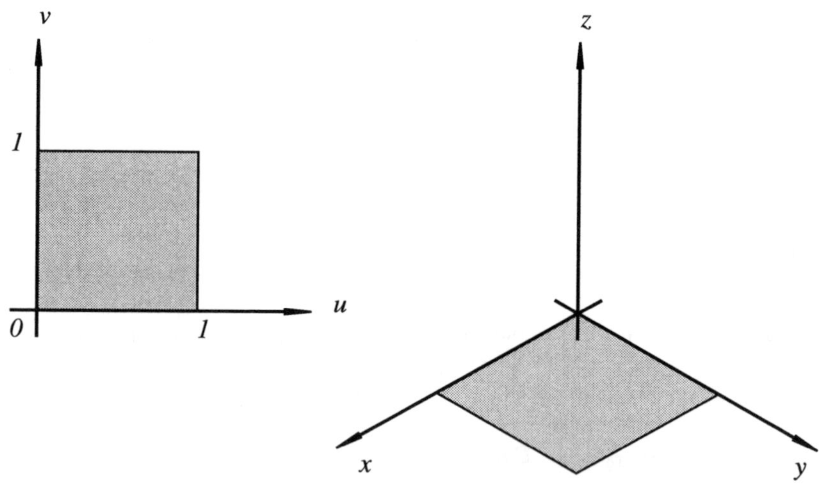

Figure 17.

Example. The parametric equation

$$\mathbf{R}(u, v) = u\mathbf{i} + v\mathbf{j}, \quad 0 \le u, v \le 1,$$

where $\mathbf{i}$, $\mathbf{j}$ are unit vectors of the coordinate axes Ox and Oy, describes a unit square on the coordinate plane Oxy (Figure 17). The twist vector is identically equal to zero

$$\mathbf{R}_{uv}(u,v) = \mathbf{O}.$$

The parametric equation

$$\mathbf{R}(u,v) = u\mathbf{i} + uv\mathbf{j}, \quad 0 \le u, v \le 1$$

describes the same square but the twist vector

$$\mathbf{R}_{uv}(u,v) = \mathbf{j} \ne \mathbf{O}.$$

In conclusion, note a useful relation

$$M = (\mathbf{R}_{uv}\mathbf{N})$$

connecting the twist vector of the regular surface with one of its second fundamental form coefficients.

4.2 Bézier surfaces

4.2.1 Parametric equations of an elementary Bézier surface

For a given array of vertices

$$\mathbf{P} = \{\mathbf{P}_{ij}, \quad i = 0, 1, \ldots, m, \quad j = 0, 1, \ldots, n\}$$

an (elementary) Bézier surface is defined by the vector equation of the following form

$$\mathbf{R}(u,v) = \sum_{i=0}^{m} \sum_{j=0}^{n} B_i^n(u) B_j^m(v) \mathbf{P}_{ij}, \quad 0 \le u, v \le 1,$$

where

$$B_i^m(u) = \binom{m}{i} u^i (1-u)^{m-i}, \quad B_j^n(v) = \binom{n}{j} v^j (1-v)^{n-j}$$

are the Bernstein polynomials.

A **matrix form** of the parametric equation describing the Bézier surface

$$\begin{pmatrix} x(u,v) \\ y(u,v) \\ z(u,v) \end{pmatrix} = \begin{pmatrix} u^0 & \ldots & u^m \end{pmatrix} \mathbf{M}^T \begin{pmatrix} \mathbf{P}_{00} & \ldots & \mathbf{P}_{0n} \\ \ldots & \ldots & \ldots \\ \mathbf{P}_{m0} & \ldots & \mathbf{P}_{mn} \end{pmatrix} \mathbf{M} \begin{pmatrix} v^0 \\ \vdots \\ v^n \end{pmatrix},$$

$$0 \le u, v \le 1,$$

where $\mathbf{M} = (\mu_{ij})$ and $\mathbf{N} = (\nu_{ij})$ are square matrices of the orders m and n, respectively,

$$\mu_{ij} = (-1)^{j-i}\binom{m}{j}\binom{j}{i}, \quad \nu_{ij} = (-1)^{j-i}\binom{n}{j}\binom{j}{i}.$$

If the parameter domain is the rectangular

$$R = \{(u,v)|a \le u \le b, c \le v \le d\},$$

the equation of the Bézier surface may be written in the following form:

$$\mathbf{R}(u,v) = \sum_{i=0}^{m}\sum_{j=0}^{n} B_i^n\left(\frac{u-a}{b-a}\right) B_j^m\left(\frac{v-c}{d-c}\right)\mathbf{P}_{ij}.$$

Important special case. When $m = n = 3$, we have an elementary *bicubic* Bézier surface defined by sixteen vertices

$$\begin{array}{cccc} \mathbf{P}_{00}, & \mathbf{P}_{01}, & \mathbf{P}_{02}, & \mathbf{P}_{03}, \\ \mathbf{P}_{10}, & \mathbf{P}_{11}, & \mathbf{P}_{12}, & \mathbf{P}_{13}, \\ \mathbf{P}_{20}, & \mathbf{P}_{21}, & \mathbf{P}_{22}, & \mathbf{P}_{23}, \\ \mathbf{P}_{30}, & \mathbf{P}_{31}, & \mathbf{P}_{32}, & \mathbf{P}_{33}. \end{array}$$

With the Bernstein polynomials this surface may be given as follows

$$\mathbf{R}(u,v) = \sum_{i=0}^{3} B_i^3(u)\left(\sum_{j=0}^{3} B_j^3(v)\mathbf{P}_{ij}\right), \quad 0 \le u, v \le 1$$

or, in the matrix form,

$$\mathbf{R}(u,v) =$$

$$\left(B_0^3(u)\, B_1^3(u)\, B_2^3(u)\, B_3^3(u)\right)\begin{pmatrix} \mathbf{P}_{00} & \mathbf{P}_{01} & \mathbf{P}_{02} & \mathbf{P}_{03} \\ \mathbf{P}_{10} & \mathbf{P}_{11} & \mathbf{P}_{12} & \mathbf{P}_{13} \\ \mathbf{P}_{20} & \mathbf{P}_{21} & \mathbf{P}_{22} & \mathbf{P}_{23} \\ \mathbf{P}_{30} & \mathbf{P}_{31} & \mathbf{P}_{32} & \mathbf{P}_{33} \end{pmatrix}\begin{pmatrix} B_0^3(v) \\ B_1^3(v) \\ B_2^3(v) \\ B_3^3(v) \end{pmatrix}.$$

The above formula is frequently written as follows

$$\mathbf{R}(t) = \mathbf{U}^T\mathbf{M}^T\mathbf{P}\mathbf{M}\mathbf{V}, \quad 0 \le u, v \le 1,$$

where

$$\mathbf{R}(u,v) = \begin{pmatrix} x(u,v) \\ y(u,v) \\ z(u,v) \end{pmatrix}, \qquad \mathbf{U} = \begin{pmatrix} u^0 \\ u^1 \\ u^2 \\ u^3 \end{pmatrix}, \qquad \mathbf{V} = \begin{pmatrix} v^0 \\ v^1 \\ v^2 \\ v^3 \end{pmatrix},$$

$$\mathbf{P} = \begin{pmatrix} \mathbf{P}_{00} & \mathbf{P}_{01} & \mathbf{P}_{02} & \mathbf{P}_{03} \\ \mathbf{P}_{10} & \mathbf{P}_{11} & \mathbf{P}_{12} & \mathbf{P}_{13} \\ \mathbf{P}_{20} & \mathbf{P}_{21} & \mathbf{P}_{22} & \mathbf{P}_{23} \\ \mathbf{P}_{30} & \mathbf{P}_{31} & \mathbf{P}_{32} & \mathbf{P}_{33} \end{pmatrix}, \qquad \mathbf{M} = \begin{pmatrix} 1 & -3 & 3 & 1 \\ 0 & 3 & -6 & 3 \\ 0 & 0 & 3 & -3 \\ 0 & 0 & 0 & 1 \end{pmatrix}.$$

The matrix $\mathbf{M}$ is called *a basic matrix of the bicubic Bézier surface.*

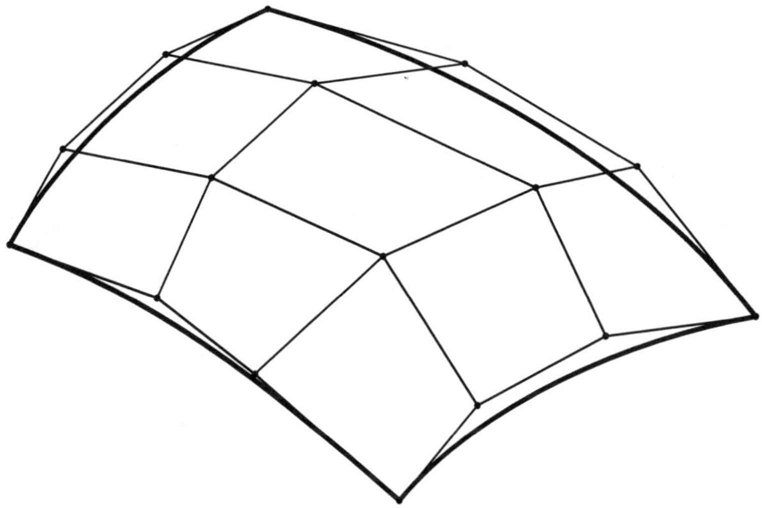

Figure 18.

4.2.2 Properties of the elementary Bézier surfaces

The properties of the elementary Bézier surfaces are the direct consequences of the correspondent properties of the elementary Bézier curves. Let us describe some of them.

Main properties of the elementary Bézier surfaces. An elementary Bézier surface generated by the array $\mathbf{P}$ possesses the following properties:

1^+ the surface is a smooth surface; in particular, the first and second derivatives of the radius-vector $\mathbf{R}(u, v)$ may be written in the form

$$\mathbf{R}_u(u, v) = \frac{\partial}{\partial u}\mathbf{R}(u, v) = m \sum_{i=0}^{m-1} \sum_{j=0}^{n} (\mathbf{P}_{i+1,j} - \mathbf{P}_{i,j}) B_i^{m-1}(u) B_j^n(v),$$

$$\mathbf{R}_v(u, v) = \frac{\partial}{\partial v}\mathbf{R}(u, v) = n \sum_{i=0}^{m} \sum_{j=0}^{n-1} (\mathbf{P}_{i,j+1} - \mathbf{P}_{i,j}) B_i^m(u) B_j^{n-1}(v);$$

if vectors $\mathbf{R}_u(u, v)$ and $\mathbf{R}_v(u, v)$ are noncollinear, a unit normal vector $\mathbf{N}(u, v)$ of the Bézier surface is defined.

2^+ the boundary curves of the elementary Bézier surface are the elementary Bézier curves of the corresponding degrees, their support polygons form a boundary of the support polyhedral surface (the support graph), in particular, the boundary curve described by radius-vector $\mathbf{R}(0, v)$ is the elementary Bézier curve of the nth degree with the support array of vertices

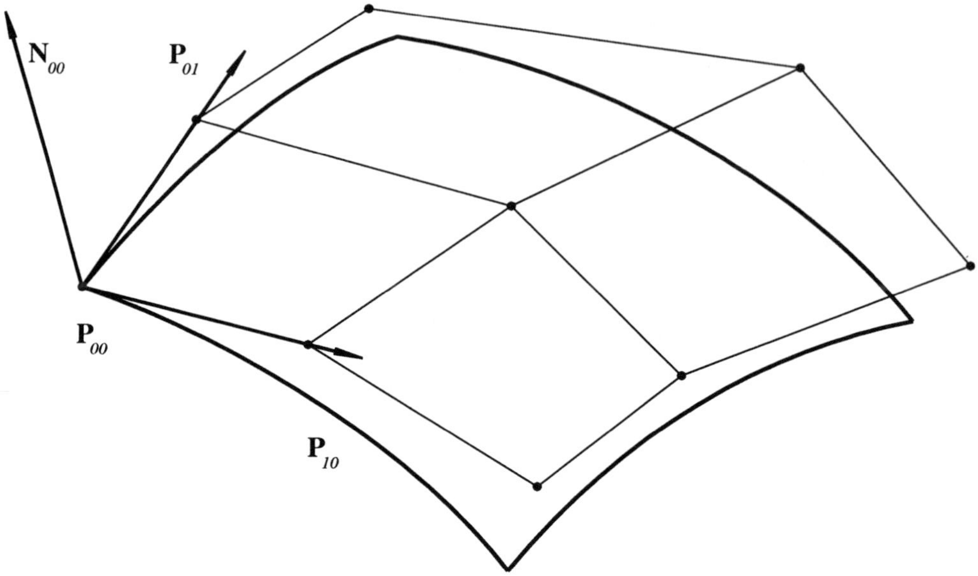

Figure 19.

$\mathbf{P}_{00}, \ldots, \mathbf{P}_{0n}$ (Figure 18); all four corner vertices of the support polyhedron $\mathbf{P}_{00}$, $\mathbf{P}_{0n}$, $\mathbf{P}_{m0}$ and $\mathbf{P}_{mn}$ lie on the Bézier surface,

$$\mathbf{R}(0,0) = \mathbf{P}_{00}, \quad \mathbf{R}(1,0) = \mathbf{P}_{m0}, \quad \mathbf{R}(0,1) = \mathbf{P}_{0n}, \quad \mathbf{R}(1,1) = \mathbf{P}_{mn},$$

and the surface is tangent to the corner faces of the control polyhedron, for example, for the corner vertex $\mathbf{P}_{00}$ ($u = v = 0$), $\mathbf{R}(0,0) = \mathbf{P}_{00}$, the following equalities

$$\mathbf{R}_u(0,0) = m(\mathbf{P}_{10} - \mathbf{P}_{00}), \quad \mathbf{R}_v(0,0) = n(\mathbf{P}_{01} - \mathbf{P}_{00})$$

(the tangent vectors of the elementary Bézier surface at the corner vertex are collinear to the edges of the corresponding boundary support polygons starting from the vertex (Figure 19),

$$\mathbf{R}_{uv}(0,0) = mn((\mathbf{P}_{11} - \mathbf{P}_{10}) - (\mathbf{P}_{01} - \mathbf{P}_{00}))$$

(the twist vector $\mathbf{R}_{uv}(0,0)$ at the corner vertex $\mathbf{P}_{00}$ is equal (up to a scalar factor) to the twist vector of the bilinear surface generated by the quadruple of vertices $\mathbf{P}_{00}$, $\mathbf{P}_{01}$, $\mathbf{P}_{10}$, $\mathbf{P}_{11}$; the vector estimates the deviation of the control vertex $\mathbf{P}_{11}$ from the tangent plane of the elementary Bézier surface at the vertex $\mathbf{P}_{00}$),

$$\mathbf{N}(0,0) = \frac{(\mathbf{P}_{10} - \mathbf{P}_{00}) \wedge (\mathbf{P}_{01} - \mathbf{P}_{00})}{|\,(\mathbf{P}_{10} - \mathbf{P}_{00}) \wedge (\mathbf{P}_{01} - \mathbf{P}_{00})\,|}$$

(the normal vector is orthogonal to the plane of the triangle $\mathbf{P}_{00}\mathbf{P}_{01}\mathbf{P}_{10}$) (similar relationships are valid for each of the other corner vertices $\mathbf{P}_{01}$ $(u = 0, v = 1)$, $\mathbf{P}_{10}$ $(u = 1, v = 0)$, $\mathbf{P}_{11}$ $(u = v = 1)$) are true.

3^+ the elementary Bézier surface lies in the convex hull generated by the array $\mathbf{P}$.

4^+ the elementary Bézier surface is affine invariant.

5^+ the elementary Bézier surface "mimicks" the control polyhedral surface.

6^+ if the support vertices of the array $\mathbf{P}$ are coplanar, then the elementary Bézier surface defined by the array is a plane curvilinear polygon lying in the planar containing the array.

7^- the degree of the functional coefficients is directly related to the number of the control vertices and increases when it grows.

8^- with addition of new vertices to the given array, it is necessary to recalculate the parametric equation of the elementary Bézier surface.

9^- the change of one vertex in the given array causes the whole elementary Bézier surface shape to change.

10^- *a priori* information on arrangement of the elementary Bézier surface (the convex hull of the given vertices contains the surface) is too rough.

11^- the equation of the elementary Bézier surface does not contain free parameters (because of this, the surface is uniquely determined by the given array of vertices and there is no way of controlling its shape).

12^- the elementary Bézier surface is projective-noninvariant.

4.2.3 Composite Bézier surfaces

In this section we consider the construction of the composite bicubic Bézier surfaces in detail. Such surfaces are most often used in the applications. The composite Bézier surfaces of the higher degrees are constructed in a similar way.

C^0-smooth (continuous) surface S is called *a composite bicubic Bézier surface* if the surface is a union of the elementary bicubic Bézier surfaces.

Let

$$\mathbf{R} = \mathbf{R}^{(1)}(u, v), \quad (u, v) \in [0, 1] \times [0, 1]$$

and

$$\mathbf{R} = \mathbf{R}^{(2)}(u, v), \quad (u, v) \in [0, 1] \times [0, 1]$$

be radius-vectors of two elementary bicubic Bézier surfaces $S^{(1)}$ and $S^{(2)}$, respectively, generated by the arrays

$$\mathbf{P}^{(1)} = \{\mathbf{P}_{ij}^{(1)}, \quad i = 0, 1, \ldots, m, \quad j = 0, 1, \ldots, n\}$$

and

$$\mathbf{P}^{(2)} = \{\mathbf{P}_{ij}^{(2)}, \quad i = 0, 1, \ldots, m, \quad j = 0, 1, \ldots, n\}$$

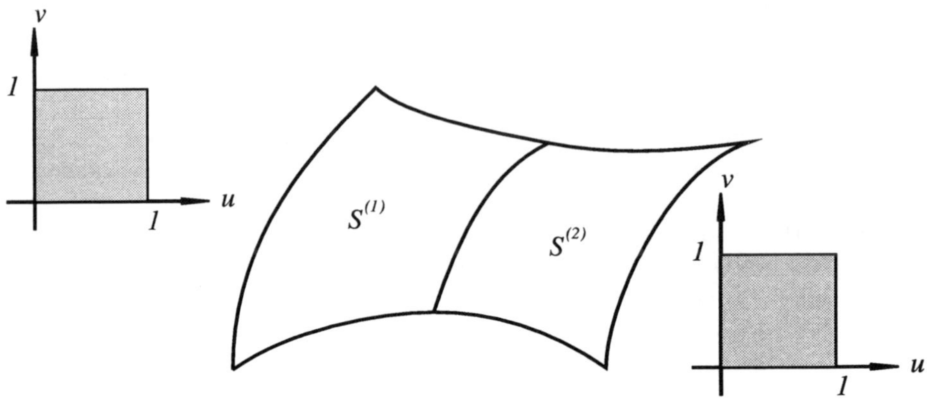

Figure 20.

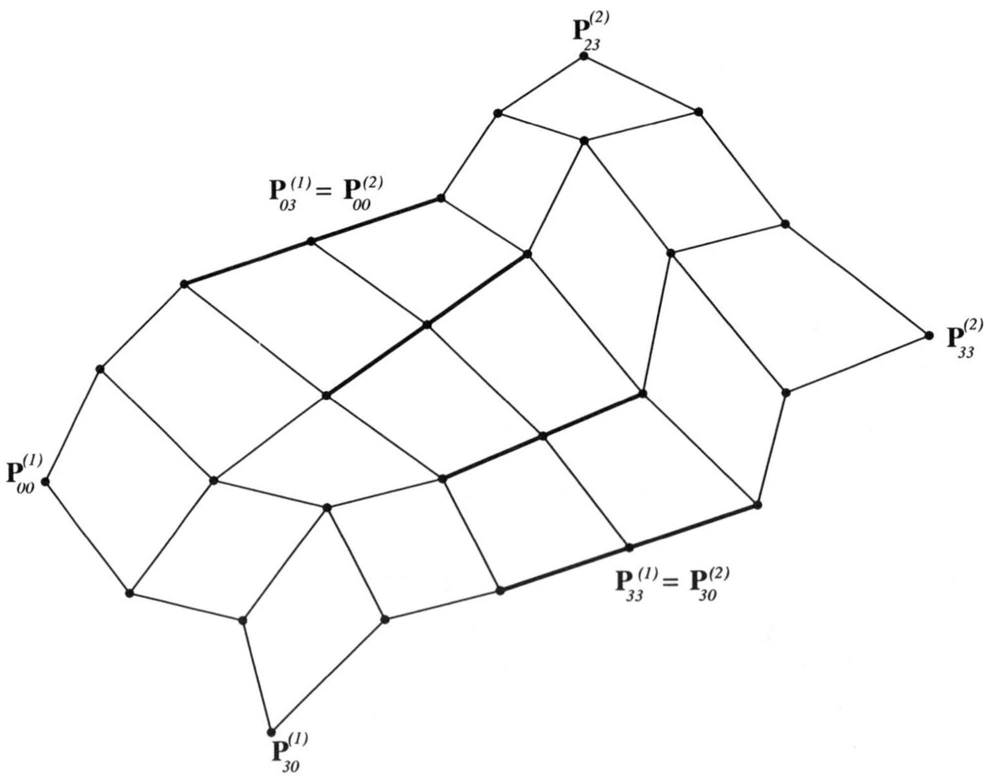

Figure 21.

such that

$$\mathbf{P}^{(1)}_{mj} = \mathbf{P}^{(2)}_{0j}, \quad j = 0, \ldots, n.$$

The last condition means that the surfaces $S^{(1)}$ and $S^{(2)}$ have a common boundary curve

$$\mathbf{R}^{(1)}(1, v) = \mathbf{R}^{(2)}(0, v), \quad 0 \leq v \leq 1$$

(Figure 20).

C^1-smoothness of the composite surface $S = S^{(1)} \cup S^{(2)}$ is guaranteed by the collinearity of all triples

$$\mathbf{P}^{(1)}_{m-1,j}, \quad \mathbf{P}^{(1)}_{mj} = \mathbf{P}^{(2)}_{0j}, \quad \mathbf{P}^{(2)}_{1j}$$

and by the fulfillment of the condition

$$\frac{| \mathbf{P}^{(1)}_{mj} - \mathbf{P}^{(1)}_{m-1,j} |}{| \mathbf{P}^{(2)}_{1j} - \mathbf{P}^{(2)}_{0,j} |} = \text{const}$$

for all j (Figure 21). The above conditions imply that the normal of the composite surface (or, what is the same, its tangent plane) changes continuously in going through the common boundary curve of the joining fragments.

4.2.4 Rational Bézier surfaces

For a given array of vertices

$$\mathbf{P} = \{\mathbf{P}_{ij}, \quad i = 0, 1, \ldots, m, \quad j = 0, 1, \ldots, n\}$$

an (elementary) rational Bézier surface is defined by an equation of the following form

$$\mathbf{R}(u, v) = \frac{\sum_{i=0}^{m} \sum_{j=0}^{n} w_{ij} B_i^m(u) B_j^n(v) \mathbf{P}_{ij}}{\sum_{i=0}^{m} \sum_{j=0}^{n} w_{ij} B_i^m(u) B_j^n(v)}, \quad 0 \leq u, v \leq 1,$$

where $B_i^m(u)$ and $B_j^n(v)$ are the Bernstein polynomials.

The nonnegative numbers w_{ij}, the sum of which is positive, are called *weights.* When all the weights w_{ij} are the same we have a standard elementary Bézier surface.

Important special case. When $m = n = 3$, we have *an elementary rational bicubic Bézier surface* generated by the array of sixteen vertices

$$\begin{array}{llll} \mathbf{P}_{00}, & \mathbf{P}_{01}, & \mathbf{P}_{02}, & \mathbf{P}_{03}, \\ \mathbf{P}_{10}, & \mathbf{P}_{11}, & \mathbf{P}_{12}, & \mathbf{P}_{13}, \\ \mathbf{P}_{20}, & \mathbf{P}_{21}, & \mathbf{P}_{22}, & \mathbf{P}_{23}, \\ \mathbf{P}_{30}, & \mathbf{P}_{31}, & \mathbf{P}_{32}, & \mathbf{P}_{33} \end{array}$$

and described by equation

$$\mathbf{R}(u,v) = \frac{1}{w(u,v)} \sum_{i=0}^{3} \sum_{j=0}^{3} w_{ij} B_i^3(u) B_j^3(v) \mathbf{P}_{ij}, \quad 0 \le u, v \le 1,$$

where

$$w(u,v) = \sum_{i=0}^{3} \sum_{j=0}^{3} w_{ij} B_i^3(u) B_j^3(v).$$

Properties of the rational bicubic Bézier surfaces. An elementary rational bicubic Bézier surface generated by array $\mathbf{P}$ possesses the following properties:

1^+ the surface is smooth.

2^+ the boundary curves of the elementary rational bicubic Bézier surface are the elementary rational cubic Bézier curves (each of the curves is controlled by its *own* set of the weight coefficients); their control polygons form the boundary of the control polyhedral surface (support graph), in particular, the boundary curve described by radius-vector $\mathbf{R}(u,v)$ is the elementary rational cubic Bézier curve with the control array of vertices $\mathbf{P}_{00}$, $\mathbf{P}_{01}$, $\mathbf{P}_{02}$, $\mathbf{P}_{03}$; the quadruple of the corner vertices of the support polyhedron $\mathbf{P}_{00}$, $\mathbf{P}_{03}$, $\mathbf{P}_{30}$, $\mathbf{P}_{33}$ lies on the Bézier surface,

$$\mathbf{R}(0,0) = \mathbf{P}_{00}, \quad \mathbf{R}(1,0) = \mathbf{P}_{30}, \quad \mathbf{R}(0,1) = \mathbf{P}_{03}, \quad \mathbf{R}(1,1) = \mathbf{P}_{33}$$

and the surface is tangent to the corner faces of the support polyhedron, for example, for the corner vertex $\mathbf{P}_{00}$ ($u = v = 0$), $\mathbf{R}(0,0) = \mathbf{P}_{00}$, the equalities containing three weight ratios

$$\frac{w_{01}}{w_{00}}, \quad \frac{w_{10}}{w_{00}}, \quad \frac{w_{11}}{w_{00}} :$$

$$\mathbf{R}_u(0,0) = 3\frac{w_{10}}{w_{00}}(\mathbf{P}_{10} - \mathbf{P}_{00}), \quad \mathbf{R}_v(0,0) = 3\frac{w_{01}}{w_{00}}(\mathbf{P}_{01} - \mathbf{P}_{00})$$

(the tangent vectors of the elementary rational bicubic Bézier surface at the corner vertex are collinear to the edges of the corresponding boundary support polygons starting from the vertex),

$$\mathbf{R}_{uv}(0,0) = 9\left(\frac{w_{11}}{w_{00}}(\mathbf{P}_{11} - \mathbf{P}_{00}) + \frac{w_{01}w_{00}}{w_{00}^2}(2\mathbf{P}_{00} - \mathbf{P}_{01} - \mathbf{P}_{10})\right)$$

(the twist vector $\mathbf{R}_{uv}(0,0)$ at the corner vertex $\mathbf{P}_{00}$ estimates the deviation of the control vertex $\mathbf{P}_{11}$ from the tangent plane of the elementary rational bicubic Bézier surface at the vertex $\mathbf{P}_{00}$) (for each of the other corner vertices $\mathbf{P}_{01}$ ($u = 0, v = 1$), $\mathbf{P}_{10}$ ($u = 1, v = 0$), $\mathbf{P}_{11}$ ($u = v = 1$) the similar relations are true) are valid.

3^+ the elementary rational bicubic Bézier surface lies in the convex hull generated by the array $\mathbf{P}$.

4^+ the elementary rational bicubic Bézier surface is affine invariant.

5^+ the elementary rational bicubic Bézier surface "mimicks" the control polyhedral surface.

6^+ if the support vertices of the array **P** are coplanar, then the elementary Bézier surface defined by the array is a plane curvilinear polygon lying in the planar containing the array.

7^+ the behavior of the rational bicubic Bézier surface is determined by the array of vertices as well as by the set of the free parameters — the weights w_{ij}, *the shape parameters*; for a given array of vertices the shape of the rational cubic Bézier surface may be controlled using the weight factors.

8^- the change of one vertex in the given array causes the whole elementary Bézier surface shape to change.

9^- *a priori* information on arrangement of the elementary Bézier surface (the convex hull of the given vertices contains the surface) is too rough.

10^+ the elementary Bézier surface is projective invariant.

Remark. In the construction of the composite rational bicubic Bézier surface it should be remembered that

1) the generated array

$$\mathbf{P} = \{\mathbf{P}_{ij}, \quad i = 0, 1, \ldots, m, \quad j = 0, 1, \ldots, n\}$$

can not be arbitrary: numbers m and n must be in the form $m = 3k + 1$, $n = 3l + 1$ (the condition may be easily satisfied by addition of new vertices to the initial array), and

2) although the choice of the weights affects the shape of the composite surface, generally the surface is only continuous, and the joining curves of the elementary fragments are its edges.

4.2.5 Implementation

The following C function is an implementation of the algorithm described above.

Function syntax:

```
bezier_surface( p, seg_x, seg_y, div_x, div_y, output )
```

Function parameters are:

```
Point **p          (in)   - spline control net
                            points Pij, i = 0...(m-1),
                                        j = 0...(n-1)
                            where m = 3*seg_x+1, n = 3*seg_y+1
int seg_x,seg_y (in)   - number of curve's segments
                            (for u and for v)
int div_x          (in)   - number of links in "u-polyline"
int div_y          (in)   - number of links in "v-polyline"
```

```
 Point **output  (out) - spline approximating net
                         points Qij, i = 0...seg_x*div_x
                                     j = 0...seg_y*div_y
```

The example below shows how to use this function to build a spline surface:

```
#include "lib.h"

/* Imported function declaration */
void bezier_surface( Point **p, int seg_x, int seg_y,
                     int div_x, int div_y, Point **output ) ;

/* Number of segments */
#define SEG   (2)

/* number of points in control graph */
#define  M    ( SEG*3+1 )
#define DIV   (5)

/* number of points in approximating net */
#define  N   (SEG*DIV+1)

/* Control graph initialization */
Point p[ M ][ M ] ;
Point p30 = {  .5, -.5,  0 } ;
Point p22 = { .33, 1, -.33 } ;
Point p06 = { 0.0, 1, -1.0 } ;
Point p60 = {   1, -1,  0  } ;
Point p54 = { .83, 2, -.67 } ;
Point p36 = {  .5, 5, -1   } ;

/* auxiliary variable */
Point tt = { 0, 0, 0 } ;

/* Declaration of output arrays */
Point oo[ N ][ N ] ;

Point EyePos = { .5, 6, 2 } ;
Point LookAt = { .5, 0, -3 } ;
Vec Up = { 0, 1, -1 } ;
Flt Alpha = .7 ;

void main( void )
{
    int i, j ;
```

```
    /* Point array fill */
    for( i = 0; i < M; i ++ )
        for( j = 0; j < M; j ++ )
        {
            tt[ 0 ] = (Flt )i / ((Flt )( M - 1 )) ;
            tt[ 2 ] = - (Flt )j / ((Flt )( M - 1 )) ;
            VecCopy( tt, p[ i ][ j ] ) ;
        }
    VecCopy( p30, p[ 3 ][ 0 ] ) ;
    VecCopy( p22, p[ 2 ][ 2 ] ) ;
    VecCopy( p06, p[ 0 ][ 6 ] ) ;
    VecCopy( p60, p[ 6 ][ 0 ] ) ;
    VecCopy( p54, p[ 5 ][ 4 ] ) ;
    VecCopy( p36, p[ 3 ][ 6 ] ) ;

    /* calculate spline */
   bezier_surface( (Point *)p, SEG, SEG,
                   DIV, DIV, (Point **)oo ) ;

    /* initialize graphics mode */
    SetVideoMode() ;

    /* initialize 3D graphics */
    Init3D( EyePos, LookAt, Up, Alpha ) ;

    /* display built spline */
    PlotSurface( (Point **)oo, M, M, 1 ) ;
}
```

The source of the program used may be found on the applied diskette in `bezier.c` file of the subdirectory `SURFACES`. The detailed information on the file names and their contents is presented in `readme.txt` file of directory `SPLINES` and in Appendix B.

4.3 B-spline surfaces

4.3.1 Parametric equations of an elementary bicubic B-spline surface

For a given array

$$\begin{matrix} \mathbf{P}_{00}, & \mathbf{P}_{01}, & \mathbf{P}_{02}, & \mathbf{P}_{03}, \\ \mathbf{P}_{10}, & \mathbf{P}_{11}, & \mathbf{P}_{12}, & \mathbf{P}_{13}, \\ \mathbf{P}_{20}, & \mathbf{P}_{21}, & \mathbf{P}_{22}, & \mathbf{P}_{23}, \\ \mathbf{P}_{30}, & \mathbf{P}_{31}, & \mathbf{P}_{32}, & \mathbf{P}_{33} \end{matrix}$$

an (elementary) bicubic B-spline surface is defined by the parametric vector equation of the following form

$$\mathbf{R}(u,v) = \sum_{i=0}^{3}\sum_{j=0}^{3} n_i(u)n_j(v)\mathbf{P}_{ij}, \quad 0 \le u, v \le 1$$

in which the functional coefficients $n_i(u)$ and $n_j(v)$ are described with the formulas

$$n_0(u) = \frac{1}{6}(1-u)^3, \quad n_1(u) = \frac{1}{6}(3u^3 - 6u^2 + 4),$$

$$n_2(u) = \frac{1}{6}(-3u^3 + 3u^2 + 3u + 1), \quad n_3(u) = \frac{u^3}{6}$$

(the formulas for polynomials $n_j(v)$ resulted from the above formulas through changing variable u by variable v).

Remark. The following form of the parametric equation of the bicubic B-spline surface

$$\mathbf{R}(u,v) = \sum_{i=0}^{3} n_i(u)\left(\sum_{j=0}^{3} n_j(v)\mathbf{P}_{ij}\right)$$

underlines an intimate connection between the cubic B-spline curves and bicubic B-spline surfaces.

A **matrix form** of the parametric equation describing the elementary bicubic B-spline surface,

$$\mathbf{R}(u,v) =$$

$$(n_0(u)n_1(u)n_2(u)n_3(u)) \begin{pmatrix} \mathbf{P}_{00} & \mathbf{P}_{01} & \mathbf{P}_{02} & \mathbf{P}_{03} \\ \mathbf{P}_{10} & \mathbf{P}_{11} & \mathbf{P}_{12} & \mathbf{P}_{13} \\ \mathbf{P}_{20} & \mathbf{P}_{21} & \mathbf{P}_{22} & \mathbf{P}_{23} \\ \mathbf{P}_{30} & \mathbf{P}_{31} & \mathbf{P}_{32} & \mathbf{P}_{33} \end{pmatrix} \begin{pmatrix} n_0(v) \\ n_1(v) \\ n_2(v) \\ n_3(v) \end{pmatrix},$$

$$0 \le u, v \le 1.$$

The last formula may be written as follows

$$\mathbf{R}(t) = \mathbf{U}^T\mathbf{M}^T\mathbf{P}\mathbf{M}\mathbf{V}, \quad 0 \le u, v \le 1,$$

where

$$\mathbf{R}(u,v) = \begin{pmatrix} x(u,v) \\ y(u,v) \\ z(u,v) \end{pmatrix}, \quad \mathbf{U} = \begin{pmatrix} u^0 \\ u^1 \\ u^2 \\ u^3 \end{pmatrix}, \quad \mathbf{V} = \begin{pmatrix} v^0 \\ v^1 \\ v^2 \\ v^3 \end{pmatrix},$$

$$\mathbf{P} = \begin{pmatrix} \mathbf{P}_{00} & \mathbf{P}_{01} & \mathbf{P}_{02} & \mathbf{P}_{03} \\ \mathbf{P}_{10} & \mathbf{P}_{11} & \mathbf{P}_{12} & \mathbf{P}_{13} \\ \mathbf{P}_{20} & \mathbf{P}_{21} & \mathbf{P}_{22} & \mathbf{P}_{23} \\ \mathbf{P}_{30} & \mathbf{P}_{31} & \mathbf{P}_{32} & \mathbf{P}_{33} \end{pmatrix},$$

$$\mathbf{M} = \begin{pmatrix} 1 & -3 & -3 & -1 \\ 4 & 0 & -6 & 3 \\ 1 & 3 & 3 & -3 \\ 0 & 0 & 0 & 1 \end{pmatrix}.$$

Matrix $\mathbf{M}$ is called *a basic matrix of bicubic B-spline surface.*

4.3.2 Properties of the elementary bicubic B-spline surfaces

Properties of the elementary bicubic B-spline surfaces are the direct consequences of the corresponding properties of the cubic B-spline curves. Let us describe some of them.

Main properties of the elementary bicubic B-spline surfaces. An elementary bicubic B-spline surface generated by the array $\mathbf{P}$ possesses the following properties:

1^+ the surface is smooth.

2^+ as a rule, the surface does not interpolate the vertices of the given array.

3^+ the boundary curves of the elementary bicubic B-spline surface are the elementary cubic B-spline curves; their support polygons form a boundary of the support polyhedral surfaces (the support graph), in particular, the boundary curve described by radius-vector $\mathbf{R}(0, v)$ is the elementary cubic B-spline curve with the support array of the vertices $\mathbf{P}_{00}$, $\mathbf{P}_{01}$, $\mathbf{P}_{02}$, $\mathbf{P}_{03}$.

4^+ the elementary bicubic B-spline surface lies in the convex hull generated by the given array $\mathbf{P}$ consisting of sixteen vertices; the property is the consequence of the fact that nonnegative functional coefficients $n_i(u)$ and $n_j(v)$ $(i, j = 0, 1, 2, 3)$ form a partition of unity,

$$\sum_{i=0}^{3}\sum_{j=0}^{3} n_i(u)n_j(v) = 1.$$

5^+ the elementary bicubic B-spline surface is affine invariant.

6^+ the elementary bicubic B-spline surface "mimicks" the control polyhedral surface.

7^+ if the support vertices of the array $\mathbf{P}$ are coplanar, then the elementary bicubic B-spline surface defined by the array is a planar curvilinear quadrangle lying in the plane containing the array.

8^- the change of one vertex in the given array causes the whole elementary bicubic B-spline surface shape to change.

9^- *a priori* information on arrangement of the elementary bicubic B-spline surface (the convex hull of the given vertices contains the surface) is too rough.

10^+ the equation of the elementary bicubic B-spline surface does not contain free parameters (because of this, the surface is uniquely determined by the given array of vertices and there is no way of controlling its shape).

10^- the elementary bicubic B-spline surface is projective-noninvariant.

4.3.3 Composite bicubic B-spline surfaces

C^0-smooth (continuous) surface S is called *a composite bicubic B-spline surface* defined by the array of $(m+1)(n+1)$ vertices

$$\mathbf{P} = \{\mathbf{P}_{ij}, \quad i = 0,1,\ldots,m, \quad j = 0,1,\ldots,n\} \quad m \geq 3, \quad n \geq 3$$

if the surface is a union of $(m-2)(n-2)$ elementary bicubic B-spline surfaces $S^{(1,1)}$, ..., $S^{(m-2,n-2)}$,

$$S = \cup_{i=1}^{m-2} \cup_{j=1}^{n-2} S^{(i,j)},$$

where (i,j)th surface $S^{(i,j)}$ is specified by the parametric equation of the form

$$\mathbf{R}^{(i,j)}(u,v) = \sum_{k=0}^{3} \sum_{l=0}^{3} n_k(u) n_l(v) \mathbf{P}_{k-1+i,l-1+j},$$

$$0 \leq u,v \leq 1, \quad 1 \leq i \leq m-2, \quad 1 \leq j \leq n-2$$

or, in the matrix form,

$$\mathbf{R}^{(i,j)}(u,v) = \mathbf{U}^T \mathbf{M}^T \begin{pmatrix} \mathbf{P}_{i-1,j-1} & \mathbf{P}_{i,j-1} & \mathbf{P}_{i+1,j-1} & \mathbf{P}_{i+2,j-1} \\ \mathbf{P}_{i-1,j} & \mathbf{P}_{i,j} & \mathbf{P}_{i+1,j} & \mathbf{P}_{i+2,j} \\ \mathbf{P}_{i-1,j+1} & \mathbf{P}_{i,j+1} & \mathbf{P}_{i+1,j+1} & \mathbf{P}_{i+2,j+1} \\ \mathbf{P}_{i-1,j+2} & \mathbf{P}_{i,j+2} & \mathbf{P}_{i+1,j+2} & \mathbf{P}_{i+2,j+2} \end{pmatrix} \mathbf{M}\mathbf{V},$$

$$0 \leq u,v \leq 1, \quad 1 \leq i \leq m-2, \quad 1 \leq j \leq n-2,$$

where $\mathbf{M}$ is the basic matrix of the bicubic B-spline. The mutual arrangement of the elementary fragments of the composite surface is schematically presented in Figure 22.

Properties of the composite bicubic B-spline surfaces. The composite bicubic B-spline surface generated by the array $\mathbf{P}$ possesses the following properties:

1^+ the surface is a C^2-smooth surface.

2^+ as a rule, the surface does not interpolate the vertices of the given array.

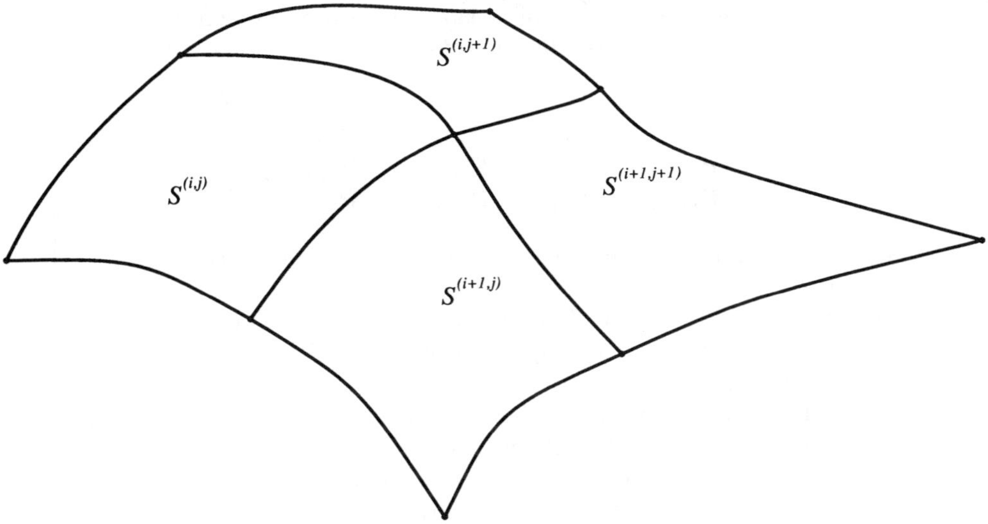

Figure 22.

(1,1)	...	(1,j)	(1,j+1)	...	(1,n-2)
...	...	...	...	...	...
(i,1)	...	(i,j)	(i,j+1)	...	(i,n-2)
(i+1,1)	...	(i+1,j)	(i+1,j+1)	...	(i+1,n-2)
...	...	...	...	...	...
(m-2,1)	...	(m-2,j)	(m-2,j+1)	...	(m-2,n-2)

Figure 23.

3^+ the boundary curves of the composite bicubic B-spline surface are the composite cubic B-spline curves; their control polygons form the boundary of the control polyhedral surface (the support graph), in particular, the boundary curve described by radius-vector $\mathbf{R}(0, v)$ is the composite cubic B-spline curve with the support array of vertices $\mathbf{P}_{00}$, ..., $\mathbf{P}_{0n}$.

4^+ the composite bicubic B-spline surface lies in the union of $(m-2) \times (n-2)$ convex hulls generated by the sets

$$\begin{matrix} \mathbf{P}_{i-1,j-1} & \mathbf{P}_{i-1,j} & \mathbf{P}_{i-1,j+1} & \mathbf{P}_{i-1,j+2} \\ \mathbf{P}_{i,j-1} & \mathbf{P}_{i,j} & \mathbf{P}_{i,j+1} & \mathbf{P}_{i,j+2} \\ \mathbf{P}_{i+1,j-1} & \mathbf{P}_{i+1,j} & \mathbf{P}_{i+1,j+1} & \mathbf{P}_{i+1,j+2} \\ \mathbf{P}_{i+2,j-1} & \mathbf{P}_{i+2,j} & \mathbf{P}_{i+2,j+1} & \mathbf{P}_{i+2,j+2} \end{matrix}$$

(a schematic partition of the composite bicubic B-spline surface is plotted in Figure 23).

5^+ the composite bicubic B-spline surface is affine invariant.

6^+ the composite bicubic B-spline surface "mimicks" the control polyhedral surface.

7^+ if the support vertices of the array $\mathbf{P}$ are coplanar, then the composite bicubic B-spline surface defined by the array is a planar curvilinear quadrangle lying in the plane containing the array.

8^+ the change of one vertex in the given array causes the surface to change in a region near the changed data only: when vertex $\mathbf{P}_{ij}$ is changed we need to recalculate the parametric equations of sixteen neighboring elementary surfaces

$$\begin{matrix} S^{(i-2,j-2)}, & S^{(i-2,j-1)}, & S^{(i-2,j)}, & S^{(i-2,j+1)}, \\ S^{(i-1,j-2)}, & S^{(i-1,j-1)}, & S^{(i-1,j)}, & S^{(i-1,j+1)}, \\ S^{(i-2,j-2)}, & S^{(i-2,j-1)}, & S^{(i-2,j)}, & S^{(i-2,j+1)}, \\ S^{(i+1,j-2)}, & S^{(i+1,j-1)}, & S^{(i+1,j)}, & S^{(i+1,j+1)}. \end{matrix}$$

9^- the composite bicubic B-spline surface is projective-noninvariant.

10^+ the equations of the composite bicubic B-spline surface do not contain free parameters (because of this, the surface is uniquely determined by the given array of vertices and there is no way of controlling its shape).

Remark. A mutual arrangement of the vertices in the array may be absolutely arbitrary (in particular, some of them may coincide), but it should be remembered that in such a situation the surface may lose its regularity. However, if the numbers of the coinciding vertices differ greatly, regularity is not necessarily lost.

Some simplest cases of the vertex coincidence are considered below.

4.3.4 Multiple and fantom vertices

In application arrays with boundary or corner multiple vertices are convenient to use. In this case the number of vertices taking part in creating the boundary bicubic elementary B-spline fragments is less than sixteen (without taking into account multiplicity). This has essential influence on the fragment sizes and makes it possible to have *a priori* information on the behavior of the composite surface in the vicinity of the boundary. As a rule, a composite bicubic B-spline surface is free from the vertices of the generated array. Nevertheless, one can gain the boundary curves of the new composite surface nearer to the boundary and corner vertices by trial of the auxiliary vertices. It is common to use the *multiple* and *fantom* vertices for this purpose.

A. Double vertices. Let us define $2m + 2n + 8$ new vertices setting

$$\mathbf{P}_{-1,j} = \mathbf{P}_{0j}, \quad j = 0, 1, \ldots, n, \quad \mathbf{P}_{m+1,j} = \mathbf{P}_{mj}, \quad j = 0, 1, \ldots, n,$$

$$\mathbf{P}_{i,-1} = \mathbf{P}_{i0}, \quad i = 0, 1, \ldots, m, \quad \mathbf{P}_{i,n+1} = \mathbf{P}_{in}, \quad i = 0, 1, \ldots, m,$$

$$\mathbf{P}_{-1,-1} = \mathbf{P}_{00}, \mathbf{P}_{-1,n+1} = \mathbf{P}_{0n}, \mathbf{P}_{m+1,-1} = \mathbf{P}_{m0}, \mathbf{P}_{m+1,n+1} = \mathbf{P}_{mn}.$$

These vertices "border" the given array $\mathbf{P}$ and, combined with the array, form a new array

$$\mathbf{P}^* = \{\mathbf{P}_{ij}, \quad i = -1, \ldots, m+1, \ j = -1, \ldots, n+1\}$$

consisting of $(m+3)(n+3)$ vertices.

Remark. The arrays $\mathbf{P}$ and $\mathbf{P}^*$ are indistinguishable (as the point sets) since all the added vertices differ from the given ones by the numbers only (geometrically we have no new vertices).

The composite bicubic B-spline surface S^* defined by the array $\mathbf{P}^*$ consists of mn fragments $S^{(i,j)}$ with the radius-vectors calculated by formulas

$$\mathbf{R}^{(i,j)}(u, v) = \sum_{k=0}^{3}\sum_{l=0}^{3} n_k(u) n_l(v) \mathbf{P}_{k-1+i,l-1+j},$$

$$0 \le u, v \le 1, \quad i = 0, \ldots, m-1, \quad j = 0, \ldots, n-1.$$

Only $2m + 2n - 4$ of these elementary surfaces

$$S^{(0,j)}, \quad j = 0, \ldots, n-2, \qquad S^{(i,n-1)}, \quad i = 0, \ldots, m-2,$$

$$S^{(m-1,j)}, \quad j = 1, \ldots, n-1, \qquad S^{(i,0)}, \quad i = 1, \ldots, m-1$$

are new (it is clear that for constructing a new fragment we need at least one new vertex).

The mutual arrangement of the composite bicubic B-spline surfaces (old S and new S^*) is easily seen in Figure 24. The new fragments are schematically displayed in Figure 25.

The values of the twist vectors at the corner vertices of four corner fragments $S^{(0,0)}$, $S^{(0,n-1)}$, $S^{(m-1,n-1)}$, $S^{(m-1,0)}$ of the surface S^*

$$R_{uv}^{(0,0)}(0,0) = \frac{1}{4}(\mathbf{P}_{00} - \mathbf{P}_{01} + \mathbf{P}_{11} - \mathbf{P}_{10}),$$

$$R_{uv}^{(0,n-1)}(0,1) = \frac{1}{4}(\mathbf{P}_{0,n-1} - \mathbf{P}_{0n} + \mathbf{P}_{1n} - \mathbf{P}_{1,n-1}),$$

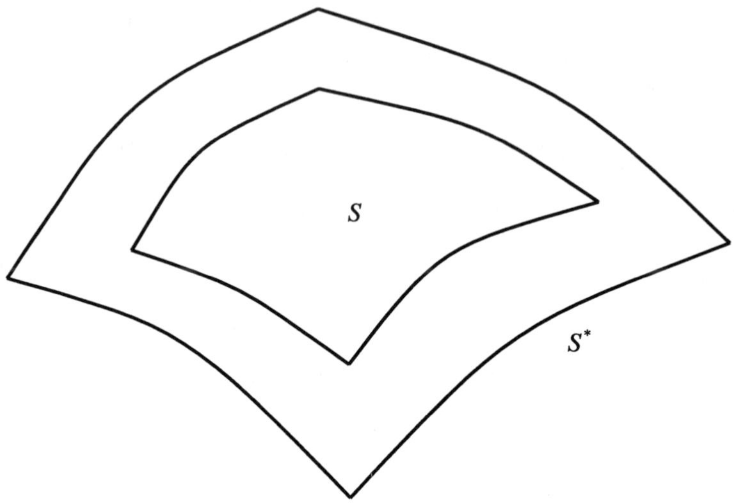

Figure 24.

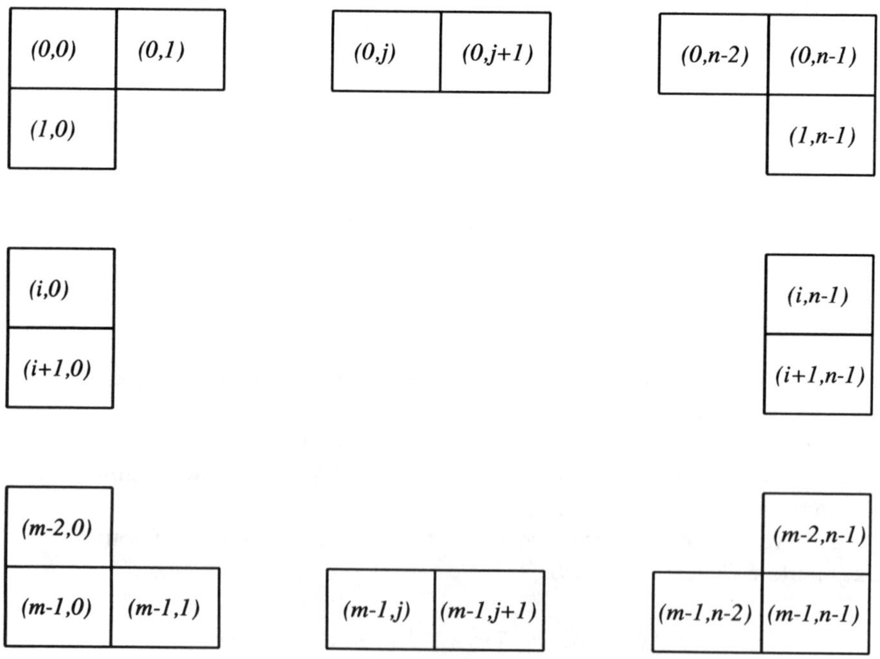

Figure 25.

$$R_{uv}^{(m-1,n-1)}(1,1) = \frac{1}{4}(\mathbf{P}_{m-1,n-1} - \mathbf{P}_{m-1,n} + \mathbf{P}_{mn} - \mathbf{P}_{m,n-1}),$$

$$R_{uv}^{(m-1,0)}(1,0) = \frac{1}{4}(\mathbf{P}_{m-1,0} - \mathbf{P}_{m-1,1} + \mathbf{P}_{11} - \mathbf{P}_{m0})$$

are equal (up to a coefficient) to the values of the twist vectors of the bilinear surfaces defined by the quadruples of vertices

$$\begin{array}{llll}
\mathbf{P}_{00}, & \mathbf{P}_{01}, & \mathbf{P}_{11}, & \mathbf{P}_{10}; \\
\mathbf{P}_{0,n-1}, & \mathbf{P}_{0n}, & \mathbf{P}_{1n}, & \mathbf{P}_{1,n-1}; \\
\mathbf{P}_{m-1,n-1}, & \mathbf{P}_{m-1,n}, & \mathbf{P}_{mn}, & \mathbf{P}_{m,n-1}; \\
\mathbf{P}_{m-1,0}, & \mathbf{P}_{m-1,1}, & \mathbf{P}_{11}, & \mathbf{P}_{m0},
\end{array}$$

respectively.

B. Triple vertices. Let us define $2m + 2n + 16$ more new vertices (in addition to $2m + 2n + 8$) setting

$$\begin{array}{ll}
\mathbf{P}_{-2,j} = \mathbf{P}_{-1,j}, j = -1, \ldots, n+1, & \mathbf{P}_{m+2,j} = \mathbf{P}_{m+1,j}, j = -1, \ldots, n+1, \\
\mathbf{P}_{i,-2} = \mathbf{P}_{i,-1}, i = -1, \ldots, m+1, & \mathbf{P}_{i,n+2} = \mathbf{P}_{i,n+1}, i = -1, \ldots, m+1,
\end{array}$$

$$\mathbf{P}_{-2,-2} = \mathbf{P}_{-1,-1}, \quad \mathbf{P}_{-2,n+2} = \mathbf{P}_{-1,n+1},$$

$$\mathbf{P}_{m+2,-2} = \mathbf{P}_{m+1,-1}, \quad \mathbf{P}_{m+2,n+2} = \mathbf{P}_{m+1,n+1}$$

and construct (taking into account the added vertices) $2m + 2n + 4$ new elementary surfaces

$$S^{(-1,j)}, \quad j = -1, \ldots, n-1, \qquad S^{(i,n)}, \quad i = -1, \ldots, m-1,$$

$$S^{(m,j)}, \quad j = 0, \ldots, n, \qquad S^{(i,-1)}, \quad i = 0, \ldots, m$$

with the help of the following parametric equations

$$\mathbf{R}^{(i,j)}(u,v) = \sum_{k=0}^{3}\sum_{l=0}^{3} n_k(u) n_l(v) \mathbf{P}_{k-1+i,l-1+j}, \quad 0 \le u, v \le 1.$$

As a result we obtain a new bicubic B-spline surface S^{**} consisting of $(m+2)(n+2)$ fragments (Figure 26).

The mutual arrangement of the new fragments of the surface S^{**} with respect to the old surface S^* is schematically illustrated in Figure 27.

All four corner fragments $S^{(-1,-1)}$, $S^{(-1,n)}$, $S^{(m,n)}$, $S^{(m,-1)}$ of the surface S^{**} lie in the bilinear surfaces. Because of this the twist vectors at the points of these four fragments are constant in the direction and have the following forms

$$R_{uv}^{(-1,-1)}(u,v) = 3u^2v^2(\mathbf{P}_{00} - \mathbf{P}_{01} + \mathbf{P}_{11} - \mathbf{P}_{10}),$$

$$R_{uv}^{(-1,n)}(u,v) = 3u^2v^2(\mathbf{P}_{0,n-1} - \mathbf{P}_{0n} + \mathbf{P}_{1n} - \mathbf{P}_{1,n-1}),$$

$$R_{uv}^{(m,n)}(u,v) = 3u^2v^2(\mathbf{P}_{m-1,n-1} - \mathbf{P}_{m-1,n} + \mathbf{P}_{mn} - \mathbf{P}_{m,n-1}),$$

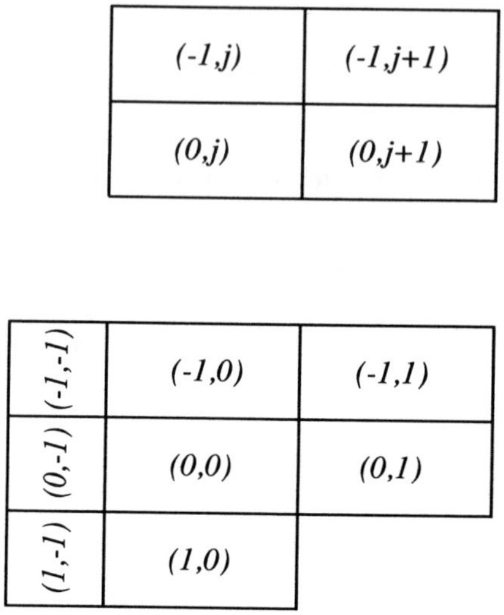

Figure 26.

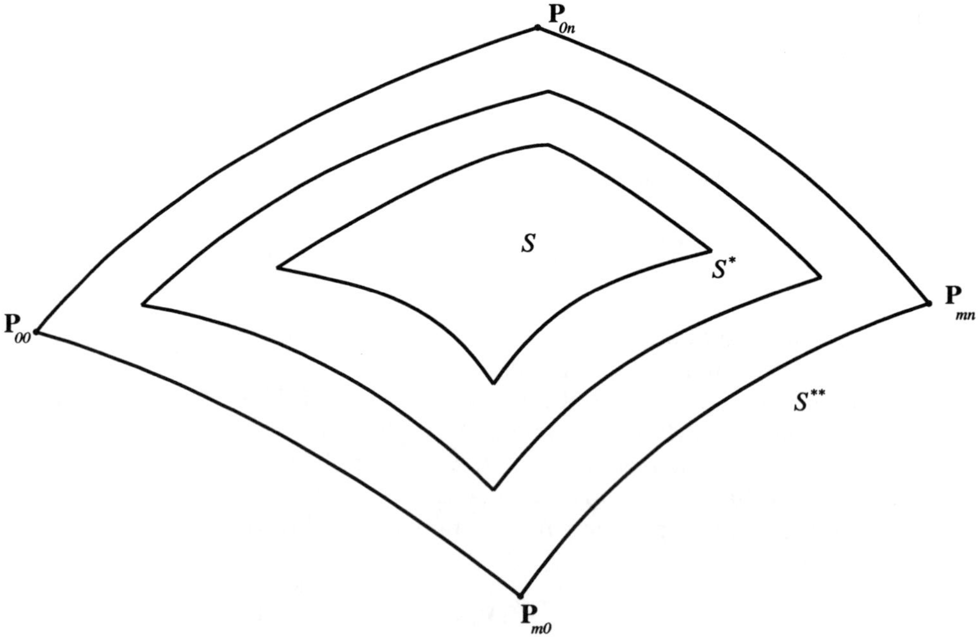

Figure 27.

$$R_{uv}^{(m,-1)}(u,v) = 3u^2v^2(\mathbf{P}_{m-1,0} - \mathbf{P}_{m-1,1} + \mathbf{P}_{11} - \mathbf{P}_{m0}),$$

respectively. In addition, all four corner vertices lie at the corners of the composite bicubic B-spline surface S^{**} constructed with the help of the triple vertices

$$\mathbf{R}^{(-1,-1)}(0,0) = \mathbf{P}_0, \quad \mathbf{R}^{(-1,n)}(0,0) = \mathbf{P}_{0n},$$

$$\mathbf{R}^{(m,n)}(0,0) = \mathbf{P}_{mn}, \quad \mathbf{R}^{(m,-1)}(0,0) = \mathbf{P}_{m0}.$$

C. Fantom vertices. By the choice of auxiliary vertices for the array

$$\mathbf{P} = \{\mathbf{P}_{ij}, \quad i = 0, \ldots, m, j = 0, \ldots, n\}$$

one can gain a fulfillment of the various conditions on the boundary of the composite surface. Usually the new (fantom) vertices are sought in the form of the linear combinations of the given vertices.

Let us consider a composite bicubic B-spline surface S^* generated by the array

$$\mathbf{P}^* = \{\mathbf{P}_{ij}, \quad i = -1, \ldots, m+1, j = -1, \ldots, n+1\}$$

of which $2m+2n+8$ vertices are selected by means of the following formulas

$$\mathbf{P}_{-1,-1} = \mathbf{P}_{0,-1} - \mathbf{P}_{1,-1} + \mathbf{P}_{0,-1},$$

$$\mathbf{P}_{m+1,n+1} = \mathbf{P}_{m+1,n} - \mathbf{P}_{m+1,n-1} + \mathbf{P}_{m+1,n},$$

$$\mathbf{P}_{-1,j} = \mathbf{P}_{0,j} - \mathbf{P}_{1,j} + \mathbf{P}_{0,j}, \quad j = 0, \ldots, n,$$

$$\mathbf{P}_{i,-1} = \mathbf{P}_{i,-1} - \mathbf{P}_{i,-1} + \mathbf{P}_{i,-1}, \quad i = 0, \ldots, m,$$

$$\mathbf{P}_{-1,n+1} = \mathbf{P}_{-1,n} - \mathbf{P}_{-1,n-1} + \mathbf{P}_{-1,n},$$

$$\mathbf{P}_{m+1,-1} = \mathbf{P}_{m,-1} - \mathbf{P}_{m-1,-1} + \mathbf{P}_{m,-1},$$

$$\mathbf{P}_{i,n+1} = \mathbf{P}_{i,n} - \mathbf{P}_{i,n-1} + \mathbf{P}_{i,n}, i = 0, \ldots, m,$$

$$\mathbf{P}_{m+1,j} = \mathbf{P}_{m,j} - \mathbf{P}_{m-1,j} + \mathbf{P}_{m,j}, j = 0, \ldots, n.$$

The composite bicubic B-spline surface S^* defined by array $\mathbf{P}^*$ possesses the following property: a curvature of the common curve of the neighboring fragments at a point lying on the boundary curve of the surface is equal to zero.

In the case under consideration, the twist vectors at the corner points of the surface S^* are equal to the twist vectors of the corresponding bilinear surfaces constructed on the edges adjacent to the vertices.

For example, the twist vector at the corner point of the fragment $S^{(0,0)}$ is equal to

$$\mathbf{P}_{00} - \mathbf{P}_{01} + \mathbf{P}_{11} - \mathbf{P}_{10}.$$

4.3.5 Rational bicubic B-spline surfaces

For a given array

$$\mathbf{P}_{ij}, \quad i = 0, \ldots, m, \quad j = 0, \ldots, n$$

an (elementary) rational bicubic B-spline surface is defined by the vector equation of the following form

$$\mathbf{R}(u,v) = \frac{\sum_{i=0}^{3}\sum_{j=0}^{3} w_{ij} n_i(u) n_j(v) \mathbf{P}_{ij}}{\sum_{i=0}^{3}\sum_{j=0}^{3} w_{ij} n_i(u) n_j(v)}, \qquad 0 \le u, v \le 1.$$

The nonnegative numbers w_{ij}, the sum of which is positive, are called *weights*. When all the weights w_{ij} are the same, we have a standard elementary bicubic B-spline surface.

Properties of the composite rational bicubic B-spline surfaces. A composite rational bicubic B-spline surface generated by array $\mathbf{P}$ possesses the following properties:

1^+ the surface is C^2-smooth surface.

2^+ as a rule, the surface does not interpolate the vertices of the given array.

3^+ the boundary curves of the composite rational bicubic B-spline surface are the composite rational cubic B-spline curves; their control polygons form the boundary of the control polyhedral surface (the support graph), in particular, the boundary curve described by radius-vector $\mathbf{R}(0, v)$ is the composite rational cubic B-spline curve with the support array of vertices $\mathbf{P}_{00}, \ldots, \mathbf{P}_{0n}$.

4^+ the composite rational bicubic B-spline surface lies in the union of $(m-2)(n-2)$ convex hulls generated by the sets of the following form

$$\{\mathbf{P}_{kl}, \quad k = i-1, i, i+1, i+2, \quad l = j-1, j, j+1, j+2\}$$

(a schematic partition of the composite rational bicubic B-spline surface is plotted in Figure 23).

5^+ the composite rational bicubic B-spline surface is affine invariant.

6^+ the composite rational bicubic B-spline surface "mimicks" the control polyhedral surface.

7^+ if the support vertices of the array $\mathbf{P}$ are coplanar, then the composite rational bicubic B-spline surface defined by the array is a planar curvilinear quadrangle lying in the plane containing the array.

8^+ the change of one vertex in the given array causes the surface to change in a region near the changed data only: when vertex $\mathbf{P}_{ij}$ is changed we need to recalculate the parametric equations of sixteen neighboring elementary surfaces

$$
\begin{array}{llll}
S^{(i-2,j-2)}, & S^{(i-2,j-1)}, & S^{(i-2,j)}, & S^{(i-2,j+1)}, \\
S^{(i-1,j-2)}, & S^{(i-1,j-1)}, & S^{(i-1,j)}, & S^{(i-1,j+1)}, \\
S^{(i,j-2)}, & S^{(i,j-1)}, & S^{(i,j)}, & S^{(i,j+1)}, \\
S^{(i+1,j-2)}, & S^{(i+1,j-1)}, & S^{(i+1,j)}, & S^{(i+1,j+1)}.
\end{array}
$$

9^+ the composite bicubic B-spline surface is projective invariant.

10^+ the behavior of the composite rational bicubic B-spline surface is determined by the array of vertices, as well as by the set of the free parameters — weights w_{ij}, *the shape parameters* (for a given array of vertices the shape of the rational bicubic B-spline surface may be controlled using the weight factors).

4.3.6 Implementation

The following C function is an implementation of the algorithm for construction of a nonuniform rational bicubic B-spline surface (NURBS).

Function syntax:

```
nurbs_surface( p, weight, uknot, vknot, m, n,
               div_x, div_y, output )
```

Function parameters are:

```
Point **p       (in)  - spline control net
                        points Pij, i = 0...(m-1),
                                    j = 0...(n-1)
Flt **w         (in)  - weights of points
                        elements Wij, i = 0...(m-1),
                                      j = 0...(n-1)
Flt uknot[]     (in)  - "u-knots",
                        elements uknot[0]...uknot[m+3]
Flt vknot[]     (in)  - "v-knots",
                        elements vknot[0]...vknot[n+3]
int m, n        (in)  - number of points in control net
int div_x       (in)  - number of links in "u-polyline"
int div_y       (in)  - number of links in "v-polyline"
Point **output (out)  - sp;ine approximating net
                        points Qij, i = 0...(m-3)*div_x
                                    j = 0...(n-3)*div_y
```

The example below shows how to use this function to build a spline surface:

```
#include "lib.h"

/* Imported function declaration */
```

```
void nurbs_surface( Point **p, Flt **weight, Flt uknot[],
                    Flt vknot[], int m, int n, int div_x,
                    int div_y, Point **output ) ;

/* number of points in control graph */
#define  M    (5)
#define DIV   (5)
/* number of points in approximating net */
#define  N    ((M-3)*DIV+1)

/* Control graph initialization */
Point p[ M ][ M ] ;
Point p11 = { .25, 1,  -.25 } ;
Point p40 = { 1.0, -.5, 0.0 } ;
Point p04 = { 0.0, 2,  -1.0 } ;
Point p33 = { .75, 2,  -.75 } ;

Flt uknot[ M + 4 ] = { 0, .1, .2, .3, .4, .5, .6, .7, .8 } ;
Flt vknot[ M + 4 ] = { 0, .1, .2, .3, .4, .5, .6, .7, .8 } ;

Flt weight[ M ][ M ] ;

/* auxiliary variable */
Point tt = { 0, 0, 0 } ;

Flt beta1 = 1.0, beta2 = 0.0 ;

/* Declaration of output arrays */
Point oo[ N ][ N ] ;

Point EyePos = { .5, 1, 1 } ;
Point LookAt = { .5, 0, -1 } ;
Vec Up = { 0, 1, 0 } ;
Flt Alpha = .7 ;

void main( void )
{
    int i, j ;

    /* Array fill */
    for( i = 0; i < M; i ++ )
        for( j = 0; j < M; j ++ )
        {
            tt[ 0 ] = (Flt )i / ((Flt )( M - 1 )) ;
            tt[ 2 ] = - (Flt )j / ((Flt )( M - 1 )) ;
```

```
                VecCopy( tt, p[ i ][ j ] ) ;

                weight[ i ][ j ] = 1.0 ;
            }
        VecCopy( p11, p[ 1 ][ 1 ] ) ;
        VecCopy( p40, p[ 4 ][ 0 ] ) ;
        VecCopy( p04, p[ 0 ][ 4 ] ) ;
        VecCopy( p33, p[ 3 ][ 3 ] ) ;

        /* calculate spline */
        nurbs_surface( (Point **)p, (Flt **)weight, uknot,
                       vknot, M, M, DIV, DIV, (Point **)oo ) ;

        /* initialize graphics mode */
        SetVideoMode() ;

        /* initialize 3D graphics */
        Init3D( EyePos, LookAt, Up, Alpha ) ;

        /* display built spline */
        PlotSurface( (Point **)oo, N, N, 1 ) ;
}
```

The source of the program used may be found on the applied diskette in `nurbs.c` file of the subdirectory `SURFACES`. The detailed information on the file names and their contents is presented in `readme.txt` file of directory `SPLINES` and in Appendix B.

4.4 Beta-spline surfaces

4.4.1 Parametric equations of an elementary Beta-spline surface

For a given array

$$\begin{array}{cccc} \mathbf{P}_{00}, & \mathbf{P}_{01}, & \mathbf{P}_{02}, & \mathbf{P}_{03}, \\ \mathbf{P}_{10}, & \mathbf{P}_{11}, & \mathbf{P}_{12}, & \mathbf{P}_{13}, \\ \mathbf{P}_{20}, & \mathbf{P}_{21}, & \mathbf{P}_{22}, & \mathbf{P}_{23}, \\ \mathbf{P}_{30}, & \mathbf{P}_{31}, & \mathbf{P}_{32}, & \mathbf{P}_{33} \end{array}$$

an (elementary) Beta-spline surface is defined by the parametric vector equation

$$\mathbf{R}(u,v) = \sum_{i=0}^{3}\sum_{j=0}^{3} b_i(u)b_j(v)\mathbf{P}_{ij}, \quad 0 \le u, v \le 1$$

with the functional coefficients $b_i(u)$ and $b_j(v)$ in the following formulas:

$$b_0(u) = \frac{2\beta_1^3}{\delta}(1-u)^3, \quad b_3(u) = \frac{2u^3}{\delta},$$

$$b_1(u) = \frac{1}{\delta}(2\beta_1^3 u(u^2 - 3u + 3) + 2\beta_1^2(u^3 - 3u^2 + 2)$$
$$+2\beta_1(u^3 - 3u + 2) + \beta_2(2u^3 - 3u^2 + 1)),$$
$$b_2(u) = \frac{1}{\delta}(2\beta_1^2 u^2(-u + 3)$$
$$+2\beta_1 u(-u^2 + 3) + \beta_2 u^2(-2u + 3) + 2(-u^3 + 1))$$

(the formulas for the polynomials $b_j(v)$ differ from the above formulas by the variables: to obtain the formulas for the polynomial coefficients $b_j(v)$ it is sufficient to replace variable u with variable v) where $\beta_1 > 0$ and $\beta_2 \geq 0$ and $\delta = 2\beta_1^3 + 4\beta_1^2 + 4\beta_1 + \beta_2 + 2$.

The numerical parameters β_1 and β_2 are called *shape parameters* of Beta-spline surface.

Remark 1. When $\beta_1 = 1$ and $\beta_2 = 0$ we have an elementary bicubic B-spline surface.

Remark 2. The following form of the parametric equation of the Beta-spline surface

$$\mathbf{R}(u,v) = \sum_{i=0}^{3} b_i(u) \left(\sum_{j=0}^{3} b_j(v)\mathbf{P}_{ij} \right)$$

underlines an intimate connection between the Beta-spline curves and the Beta-spline surfaces.

A **matrix form** of the parametric equation describing the elementary Beta-spline surface,

$$\mathbf{R}(u,v) =$$
$$\begin{pmatrix} b_0(u) \, b_1(u) \, b_2(u) \, b_3(u) \end{pmatrix} \begin{pmatrix} \mathbf{P}_{00} & \mathbf{P}_{01} & \mathbf{P}_{02} & \mathbf{P}_{03} \\ \mathbf{P}_{10} & \mathbf{P}_{11} & \mathbf{P}_{12} & \mathbf{P}_{13} \\ \mathbf{P}_{20} & \mathbf{P}_{21} & \mathbf{P}_{22} & \mathbf{P}_{23} \\ \mathbf{P}_{30} & \mathbf{P}_{31} & \mathbf{P}_{32} & \mathbf{P}_{33} \end{pmatrix} \begin{pmatrix} b_0(v) \\ b_1(v) \\ b_2(v) \\ b_3(v) \end{pmatrix},$$
$$0 \leq u, v \leq 1.$$

The last formula may be written as follows

$$\mathbf{R}(t) = \mathbf{U}^T \mathbf{M}^T \mathbf{P} \mathbf{M} \mathbf{V}, \quad 0 \leq u, v \leq 1,$$

where

$$\mathbf{R}(u,v) = \begin{pmatrix} x(u,v) \\ y(u,v) \\ z(u,v) \end{pmatrix}, \quad \mathbf{U} = \begin{pmatrix} u^0 \\ u^1 \\ u^2 \\ u^3 \end{pmatrix}, \quad \mathbf{V} = \begin{pmatrix} v^0 \\ v^1 \\ v^2 \\ v^3 \end{pmatrix},$$

$$\mathbf{P} = \begin{pmatrix} \mathbf{P}_{00} & \mathbf{P}_{01} & \mathbf{P}_{02} & \mathbf{P}_{03} \\ \mathbf{P}_{10} & \mathbf{P}_{11} & \mathbf{P}_{12} & \mathbf{P}_{13} \\ \mathbf{P}_{20} & \mathbf{P}_{21} & \mathbf{P}_{22} & \mathbf{P}_{23} \\ \mathbf{P}_{30} & \mathbf{P}_{31} & \mathbf{P}_{32} & \mathbf{P}_{33} \end{pmatrix},$$

$$\mathbf{M} = \frac{1}{\delta} \begin{pmatrix} 2\beta_1^3 & -6\beta_1^3 & 6\beta_1^3 & -2\beta_1^3 \\ 4(\beta_1^2 + \beta_1) + \beta_2 & 6(\beta_1^3 - \beta_1) & -3(2\beta_1^3 + \eta) & 2(\beta_1^3 + \theta) \\ 2 & 6\beta_1 & 3\eta & -2(\theta + 1) \\ 0 & 0 & 0 & 2 \end{pmatrix}.$$

(Here $\eta = 2\beta_1^2 + \beta_2$, $\theta = \beta_1^2 + \beta_1 + \beta_2$.)

Matrix $\mathbf{M}$ is called *a basic matrix of Beta-spline surface.*

4.4.2 Properties of the elementary Beta-spline surfaces

Properties of the elementary Beta-spline surfaces are the direct consequences of the corresponding properties of the Beta-spline curves. Let us describe some of them.

Main properties of the elementary Beta-spline surfaces. An elementary Beta-spline surface generated by array $\mathbf{P}$ possesses the following properties:

1^+ the surface is smooth.

2^+ as a rule, the surface does not interpolate the vertices of the given array.

3^+ the boundary curves of the elementary Beta-spline surface are the elementary Beta-spline curves; their support polygons form a boundary of the support polyhedral surface (the support graph). In particular, the boundary curve described by radius-vector $\mathbf{R}(0, v)$ is the elementary Beta-spline curve with the support array of the vertices $\mathbf{P}_{00}$, $\mathbf{P}_{01}$, $\mathbf{P}_{02}$, $\mathbf{P}_{03}$.

4^+ the elementary Beta-spline surface lies in the convex hull generated by the given array $\mathbf{P}$ consisting of sixteen vertices; the property is the consequence of the fact that the nonnegative functional coefficients $b_i(u)$ and $b_j(v)$ $(i, j = 0, 1, 2, 3)$ form a partition of unity

$$\sum_{i=0}^{3} \sum_{j=0}^{3} b_i(u) b_j(v) = 1.$$

5^+ the elementary Beta-spline surface is affine invariant.

6^+ the elementary Beta-spline surface "mimicks" the control polyhedral surface (the control graph).

7^+ if the support vertices of the given array $\mathbf{P}$ are coplanar, then the elementary Beta-spline surface defined by the array is a planar curvilinear quadrangle lying in the plane containing the array.

8^- the change of one vertex in the given array causes the whole elementary Beta-spline surface shape to change.

9^- *a priori* information on arrangement of the elementary Beta-spline surface (the convex hull of the given vertices contains the surface) is too rough.

10^+ the equation of the elementary Beta-spline surface contains two free parameters — *the shape parameters* β_1 and β_2; for a given array of sixteen vertices the shape of the elementary Beta-spline surface may be controlled using the shape parameters.

10^- the elementary Beta-spline surface is projective noninvariant.

4.4.3 Composite Beta-spline surfaces

C^0-smooth (continuous) surface S is called *a composite Beta-spline surface* defined by the array of $(m+1)(n+1)$ vertices

$$\mathbf{P} = \{\mathbf{P}_{ij}, \quad i = 0, 1, \ldots, m, \quad j = 0, 1, \ldots, n\} \quad m \geq 3, \quad n \geq 3,$$

if the surface is a union of $(m-2)(n-2)$ elementary Beta-spline surfaces $S^{(1,1)}, \ldots, S^{(m-2,n-2)}$,

$$S = \cup_{i=1}^{m-2} \cup_{j=1}^{n-2} S^{(i,j)},$$

where (i,j)th surface $S^{(i,j)}$ is specified by the parametric equation of the form

$$\mathbf{R}^{(i,j)}(u,v) = \sum_{k=0}^{3}\sum_{l=0}^{3} b_k(u) b_l(v) \mathbf{P}_{k-1+i,l-1+j},$$

$$0 \leq u, v \leq 1, \quad 1 \leq i \leq m-2, \quad 1 \leq j \leq n-2$$

or, in the matrix form,

$$\mathbf{R}^{(i,j)}(u,v) = \mathbf{U}^T\mathbf{M}^T \begin{pmatrix} \mathbf{P}_{i-1,j-1} & \mathbf{P}_{i,j-1} & \mathbf{P}_{i+1,j-1} & \mathbf{P}_{i+2,j-1} \\ \mathbf{P}_{i-1,j} & \mathbf{P}_{i,j} & \mathbf{P}_{i+1,j} & \mathbf{P}_{i+2,j} \\ \mathbf{P}_{i-1,j+1} & \mathbf{P}_{i,j+1} & \mathbf{P}_{i+1,j+1} & \mathbf{P}_{i+2,j+1} \\ \mathbf{P}_{i-1,j+2} & \mathbf{P}_{i,j+2} & \mathbf{P}_{i+1,j+2} & \mathbf{P}_{i+2,j+2} \end{pmatrix} \mathbf{MV},$$

$$0 \leq u, v \leq 1, \quad 1 \leq i \leq m-2, \quad 1 \leq j \leq n-2,$$

where $\mathbf{M}$ is the basic matrix of the Beta-spline.

The mutual arrangement of the elementary fragments of the composite surface is schematically presented in Figure 22.

Properties of the composite Beta-spline surfaces. The composite Beta-spline surface generated by the array $\mathbf{P}$ possesses the following properties:

1^+ the surface is G^2-continuous.

2^+ as a rule, the surface does not interpolate the vertices of the given array.

3^+ the boundary curves of the composite Beta-spline surface are the composite Beta-spline curves; their control polygons form the boundary of the control polyhedral surface (the support graph). In particular, the boundary curve described by radius-vector $\mathbf{R}(0, v)$ is the composite Beta-spline curve with the support array of vertices $\mathbf{P}_{00}, \ldots, \mathbf{P}_{0n}$.

4^+ the composite Beta-spline surface lies in the union of $(m-2)(n-2)$ convex hulls generated by the sets

$$\begin{array}{llll} \mathbf{P}_{i-1,j-1}, & \mathbf{P}_{i-1,j}, & \mathbf{P}_{i-1,j+1}, & \mathbf{P}_{i-1,j+2}, \\ \mathbf{P}_{i,j-1}, & \mathbf{P}_{i,j}, & \mathbf{P}_{i,j+1}, & \mathbf{P}_{i,j+2}, \\ \mathbf{P}_{i+1,j-1}, & \mathbf{P}_{i+1,j}, & \mathbf{P}_{i+1,j+1}, & \mathbf{P}_{i+1,j+2}, \\ \mathbf{P}_{i+2,j-1}, & \mathbf{P}_{i+2,j}, & \mathbf{P}_{i+2,j+1}, & \mathbf{P}_{i+2,j+2}. \end{array}$$

(A schematic partition of the composite Beta-spline surface is plotted in Figure 23.)

5^+ the composite Beta-spline surface is affine invariant.

6^+ the composite Beta-spline surface "mimicks" the control polyhedral surface.

7^+ if the support vertices of the array $\mathbf{P}$ are coplanar, then the composite Beta-spline surface defined by the array is a planar curvilinear quadrangle lying in the plane containing the array.

8^+ the change of one vertex in the given array causes the surface to change in a region near the changed data only: when vertex $\mathbf{P}_{i,j}$ is changed we need to recalculate the parametric equations of sixteen neighboring elementary surfaces

$$\begin{array}{llll} S^{(i-2,j-2)}, & S^{(i-2,j-1)}, & S^{(i-2,j)}, & S^{(i-2,j+1)}, \\ S^{(i-1,j-2)}, & S^{(i-1,j-1)}, & S^{(i-1,j)}, & S^{(i-1,j+1)}, \\ S^{(i,j-2)}, & S^{(i,j-1)}, & S^{(i,j)}, & S^{(i,j+1)}, \\ S^{(i+1,j-2)}, & S^{(i+1,j-1)}, & S^{(i+1,j)}, & S^{(i+1,j+1)}. \end{array}$$

9^- the composite Beta-spline surface is projective noninvariant.

10^+ the behavior of the composite Beta-spline surface is determined by the array of vertices, as well as by the pair of free parameters β_1 and β_2: for the given array of vertices the shape of the Beta-spline surface may be controlled using these shape parameters.

Remark 1. The shape parameters β_1 and β_2 need not be the same for all of the elementary fragments. In view of the mutual arrangement of the vertices in the array, the parameters may be chosen in such a way that

1) there is a pair of numbers $\beta_1^{(ij)}$ and $\beta_2^{(ij)}$ specific for every elementary fragment,

2) functions $\beta_1(u, v)$ and $\beta_2(u, v)$ depend on parameters u and v linearly on every unit square $[0, 1] \times [0, 1]$, or

3) there is a pair of functions $\beta_1^{(ij)}(u,v)$ and $\beta_2^{(ij)}(u,v)$ $(i = 1,\ldots,m-2, j = 1,\ldots,n-2)$ specific for each of the elementary fragments.

As a rule, the choice of the shape parameters β_1 and β_2 is defined by the mutual arrangement of the vertices in the array. If the quadrangles

$$\mathbf{P}_{i,j}\mathbf{P}_{i+1,j}\mathbf{P}_{i,j+1}\mathbf{P}_{i+1,j+1},$$

$$0 \le i \le m-1, \quad 0 \le j \le n-1$$

do not differ greatly in size, the choice of the parameters such that they are the same for all of the surface fragments gives sufficiently satisfactory results. If the difference of the quadrangles is very large in size, good results are obtained by the suitable choice of the variable shape parameters.

Remark 2. The mutual arrangement of the vertices in the array may be absolutely arbitrary (in particular, some of the vertices may coincide). But it should be remembered that in such a situation the surface may lose its regularity. However, if the numbers of the coinciding vertices differ greatly, regularity is not necessarily lost.

The simplest cases of the vertex coincidence are considered below.

4.4.4 Multiple and fantom vertices

In applications, arrays with multiple (coinciding) vertices which are boundary or corner are convenient to use. In this case the number of vertices taking part in creating the boundary elementary Beta-spline fragments is less than sixteen (without taking into account the multiplicity). This has an essential influence on the fragment sizes and makes it possible to have *a priori* information on the behavior of the composite surface in the vicinity of the boundary. As a rule, a composite Beta-spline surface is free from the vertices of the generated array. Nevertheless, one can gain the boundary curves of the new composite surface nearer to the boundary and corner vertices by trial with auxiliary vertices. It is common to use the *multiple* and *fantom* vertices for this purpose.

A. Double vertices. Let us define $2m + 2n + 8$ new vertices setting

$$\begin{array}{llll} \mathbf{P}_{-1,j} = \mathbf{P}_{0j}, & j = 0,1,\ldots,n, & \mathbf{P}_{m+1,j} = \mathbf{P}_{mj}, & j = 0,1,\ldots,n, \\ \mathbf{P}_{i,-1} = \mathbf{P}_{i0}, & i = 0,1,\ldots,m, & \mathbf{P}_{i,n+1} = \mathbf{P}_{in}, & i = 0,1,\ldots,m, \end{array}$$

$$\mathbf{P}_{-1,-1} = \mathbf{P}_{00}, \mathbf{P}_{-1,n+1} = \mathbf{P}_{0n}, \mathbf{P}_{m+1,-1} = \mathbf{P}_{m0}, \mathbf{P}_{m+1,n+1} = \mathbf{P}_{mn}.$$

These vertices "border" the given array $\mathbf{P}$ and, combined with the array, form a new array

$$\mathbf{P}^* = \{\mathbf{P}_{ij}, \quad i = -1,\ldots,m+1, \; j = -1,\ldots,n+1\}$$

containing $(m+3)(n+3)$ vertices.

Remark. The arrays $\mathbf{P}$ and $\mathbf{P}^*$ are indistinguishable (as the point sets) since all of the added vertices differ from the given ones only by the numbers (geometrically we have no new vertices).

The composite Beta-spline surface S^* defined by the array $\mathbf{P}^*$ consists of mn fragments $S^{(i,j)}$, the radius-vectors of which are calculated by formulas

$$\mathbf{R}^{(i,j)}(u,v) = \sum_{k=0}^{3}\sum_{l=0}^{3} b_k(u)b_l(v)\mathbf{P}_{k-1+i,l-1+j},$$

$$0 \le u, v \le 1, \quad i = 0, \ldots, m-1, \quad j = 0, \ldots, n-1.$$

Only $2m + 2n - 4$ of these elementary surfaces

$$S^{(0,j)}, \quad j = 0, \ldots, n-2, \qquad S^{(i,n-1)}, \quad i = 0, \ldots, m-2,$$

$$S^{(m-1,j)}, \quad j = 1, \ldots, n-1, \qquad S^{(i,0)}, \quad i = 1, \ldots, m-1$$

are new. (It is clear that for constructing a new fragment we need at least one new vertex.)

The mutual arrangement of the composite Beta-spline surfaces (old S and new S^*) is easily seen in Figure 24. The new fragments are schematically displayed in Figure 25.

The values of the twist vectors at the corner vertices of four corner fragments $S^{(0,0)}$, $S^{(0,n-1)}$, $S^{(m-1,n-1)}$, $S^{(m-1,0)}$ of the surface S^*

$$R_{uv}^{(0,0)}(0,0) = \frac{36}{\delta^2}(\mathbf{P}_{00} - \mathbf{P}_{01} + \mathbf{P}_{11} - \mathbf{P}_{10}),$$

$$R_{uv}^{(0,n-1)}(0,1) = \frac{36}{\delta^2}(\mathbf{P}_{0,n-1} - \mathbf{P}_{0n} + \mathbf{P}_{1n} - \mathbf{P}_{1,n-1}),$$

$$R_{uv}^{(m-1,n-1)}(1,1) = \frac{36}{\delta^2}(\mathbf{P}_{m-1,n-1} - \mathbf{P}_{m-1,n} + \mathbf{P}_{mn} - \mathbf{P}_{m,n-1}),$$

$$R_{uv}^{(m-1,0)}(1,0) = \frac{36}{\delta^2}(\mathbf{P}_{m-1,0} - \mathbf{P}_{m-1,1} + \mathbf{P}_{11} - \mathbf{P}_{m0})$$

are equal (up to the coefficients) to the values of the twist vectors of the bilinear surfaces defined by the quadruples of vertices,

$$\begin{array}{llll} \mathbf{P}_{00}, & \mathbf{P}_{01}, & \mathbf{P}_{11}, & \mathbf{P}_{10}; \\ \mathbf{P}_{0,n-1}, & \mathbf{P}_{0n}, & \mathbf{P}_{1n}, & \mathbf{P}_{1,n-1}; \\ \mathbf{P}_{m-1,n-1}, & \mathbf{P}_{m-1,n}, & \mathbf{P}_{mn}, & \mathbf{P}_{m,n-1}; \\ \mathbf{P}_{m-1,0}, & \mathbf{P}_{m0}, & \mathbf{P}_{m1}, & \mathbf{P}_{m-1,1}, \end{array}$$

respectively.

B. Triple vertices. Let us define $2m + 2n + 16$ more new vertices (in addition to $2m + 2n + 8$) setting

$$\mathbf{P}_{-2,j} = \mathbf{P}_{-1,j}, j = -1, \ldots, n+1, \quad \mathbf{P}_{m+2,j} = \mathbf{P}_{m+1,j}, j = -1, \ldots, n+1,$$
$$\mathbf{P}_{i,-2} = \mathbf{P}_{i,-1}, i = -1, \ldots, m+1, \quad \mathbf{P}_{i,n+2} = \mathbf{P}_{i,n+1}, i = -1, \ldots, m+1,$$

$$\mathbf{P}_{-2,-2} = \mathbf{P}_{-1,-1}, \quad \mathbf{P}_{-2,n+2} = \mathbf{P}_{-1,n+1},$$

$$\mathbf{P}_{m+2,-2} = \mathbf{P}_{m+1,-1}, \quad \mathbf{P}_{m+2,n+2} = \mathbf{P}_{m+1,n+1}$$

and construct (taking into account the added vertices) $2m + 2n + 4$ new elementary fragments

$$S^{(-1,j)}, \quad j = -1, \ldots, n-1, \qquad S^{(i,n)}, \quad i = -1, \ldots, m-1,$$

$$S^{(m,j)}, \quad j = 0, \ldots, n, \qquad S^{(i,-1)}, \quad i = 0, \ldots, m$$

with the help of the following parametric equations

$$\mathbf{R}^{(i,j)}(u,v) = \sum_{k=0}^{3}\sum_{l=0}^{3} b_k(u) b_l(v) \mathbf{P}_{k-1+i,l-1+j}, \quad 0 \le u, v \le 1.$$

As a result we obtain a new Beta-spline surface S^{**} consisting of $(m + 2) \times (n + 2)$ fragments (Figure 26).

The mutual arrangement of the new fragments of the surface S^{**} with respect to the old surface S^* is schematically illustrated in Figure 27.

All four corner fragments $S^{(-1,-1)}$, $S^{(-1,n)}$, $S^{(m,n)}$, $S^{(m,-1)}$ of the surface S^{**} lie in their bilinear surfaces. Because of this the twist vectors at the points of these four fragments are constant in the directions and have the following forms

$$R_{uv}^{(-1,-1)}(u,v) = \frac{36u^2v^2}{\delta}(\mathbf{P}_{00} - \mathbf{P}_{01} + \mathbf{P}_{11} - \mathbf{P}_{10}),$$

$$R_{uv}^{(-1,n)}(u,v) = \frac{36u^2v^2}{\delta}(\mathbf{P}_{0,n-1} - \mathbf{P}_{0n} + \mathbf{P}_{1n} - \mathbf{P}_{1,n-1}),$$

$$R_{uv}^{(m,n)}(u,v) = \frac{36u^2v^2}{\delta}(\mathbf{P}_{m-1,n-1} - \mathbf{P}_{m-1,n} + \mathbf{P}_{mn} - \mathbf{P}_{m,n-1}),$$

$$R_{uv}^{(m,-1)}(u,v) = \frac{36u^2v^2}{\delta}(\mathbf{P}_{m-1,0} - \mathbf{P}_{m-1,1} + \mathbf{P}_{11} - \mathbf{P}_{m0}),$$

respectively. In addition, all four corner vertices lie at the corners of the composite Beta-spline surface S^{**} constructed with the help of the triple vertices

$$\mathbf{R}^{(-1,-1)}(0,0) = \mathbf{P}_0, \quad \mathbf{R}^{(-1,n)}(0,0) = \mathbf{P}_{0n},$$

$$\mathbf{R}^{(m,n)}(0,0) = \mathbf{P}_{mn}, \quad \mathbf{R}^{(m,-1)}(0,0) = \mathbf{P}_{m0}.$$

C. Fantom vertices. By the choice of the auxiliary vertices to the array

$$\mathbf{P} = \{\mathbf{P}_{ij}, \quad i = 0, \ldots, m, j = 0, \ldots, n\}$$

one can gain a fulfillment of the various conditions on the boundary of the composite surface. Usually the new (fantom) vertices are sought in the form of the linear combinations of the given vertices.

Let us consider a composite Beta-spline surface S^* generated by the array

$$\mathbf{P}^* = \{\mathbf{P}_{ij}, \quad i = -1, \ldots, m+1, j = -1, \ldots, n+1\}$$

of which $2m+2n+8$ vertices are selected by means of the following formulas

$$\mathbf{P}_{-1,-1} = \frac{2\beta_1^2 + \beta_2}{2\beta_1^3}(\mathbf{P}_{0,-1} - \mathbf{P}_{1,-1}) + \mathbf{P}_{0,-1},$$

$$\mathbf{P}_{m+1,n+1} = \frac{2\beta_1^2 + \beta_2}{2\beta_1^3}(\mathbf{P}_{m+1,n} - \mathbf{P}_{m+1,n-1}) + \mathbf{P}_{m+1,n},$$

$$\mathbf{P}_{-1,j} = \frac{2\beta_1^2 + \beta_2}{2\beta_1^3}(\mathbf{P}_{0,j} - \mathbf{P}_{1,j}) + \mathbf{P}_{0,j}, \quad j = 0, \ldots, n,$$

$$\mathbf{P}_{i,-1} = \frac{2\beta_1^2 + \beta_2}{2\beta_1^3}(\mathbf{P}_{i,-1} - \mathbf{P}_{i,-1}) + \mathbf{P}_{i,-1}, \quad i = 0, \ldots, m,$$

$$\mathbf{P}_{-1,n+1} = \frac{2\beta_1 + \beta_2}{2}(\mathbf{P}_{-1,n} - \mathbf{P}_{-1,n-1}) + \mathbf{P}_{-1,n},$$

$$\mathbf{P}_{m+1,-1} = \frac{2\beta_1 + \beta_2}{2}(\mathbf{P}_{m,-1} - \mathbf{P}_{m-1,-1}) + \mathbf{P}_{m,-1},$$

$$\mathbf{P}_{i,n+1} = \frac{2\beta_1 + \beta_2}{2}(\mathbf{P}_{i,n} - \mathbf{P}_{i,n-1}) + \mathbf{P}_{i,n}, i = 0, \ldots, m,$$

$$\mathbf{P}_{m+1,j} = \frac{2\beta_1 + \beta_2}{2}(\mathbf{P}_{m,j} - \mathbf{P}_{m-1,j}) + \mathbf{P}_{m,j}, j = 0, \ldots, n.$$

The composite Beta-spline surface S^* defined by array $\mathbf{P}^*$ possesses the following property: a curvature of the common curve of the neighboring fragments at the point lying on the boundary curve of the surface is equal to zero.

In the case under consideration, the twist vectors at the corner points of the surface S^* are equal up to the coefficients to the twist vectors of the corresponding bilinear surfaces constructed on the edges adjacent to the vertices.

For example, the twist vector at the corner point of the fragment $S^{(0,0)}$ is equal to

$$\alpha(\mathbf{P}_{00} - \mathbf{P}_{01} + \mathbf{P}_{11} - \mathbf{P}_{10}),$$

where

$$\alpha = \frac{9}{\delta^2}(4\beta_1^4 + 8\beta_1^3 + 4\beta_1^2\beta_2 + 4\beta_1^2 + 4\beta_1\beta_2 + \beta_2^2).$$

4.4.5 Implementation

The following C function is an implementation of the algorithm described above.

Function syntax:

```
beta_surface( b1, b2, p, m, n, div_x, div_y, output )
```

Function parameters are:

```
Flt b1, b2       (in)  - beta1 and beta2
Point **p        (in)  - spline control net
                         points Pij, i = 0...(m-1),
                                     j = 0...(n-1)
int m, n         (in)  - number of points in control net
int div_x        (in)  - number of links in "u-polyline"
int div_y        (in)  - number of links in "v-polyline"
Point **output (out)  - spline approximating net
                         points Qij, i = 0...(m-3)*div_x
                                     j = 0...(n-3)*div_y
```

The example below shows how to use this function to build a spline surface:

```
#include "lib.h"

/* Imported function declaration */
void beta_surface( Flt b1, Flt b2, Point **p, int m, int n,
                   int div_x, int div_y, Point **output ) ;

/* number of points in control graph */
#define  M    (5)
#define DIV   (5)
/* number of poits in approximating net */
#define  N    ((M-3)*DIV+1)

/* Control graph initialization */
Point p[ M ][ M ] ;
Point p11 = { .25, 1,  -.25 } ;
Point p40 = { 1.0, -.5, 0.0 } ;
Point p04 = { 0.0, 2,  -1.0 } ;
Point p33 = { .75, 2,  -.75 } ;

/* auxiliary variable */
Point tt = { 0, 0, 0 } ;
```

```
Flt beta1 = 1.0, beta2 = 0.0 ;

/* Declaration of output arrays */
Point oo[ N ][ N ] ;

Point EyePos = { .5, 1, 1 } ;
Point LookAt = { .5, 0, .5 } ;
Vec Up = { 0, 1, 0 } ;
Flt Alpha = .7 ;

void main( void )
{
    int i, j ;

    /* Point array fill */
    for( i = 0; i < M; i ++ )
        for( j = 0; j < M; j ++ )
        {
            tt[ 0 ] = (Flt )i / ((Flt )( M - 1 )) ;
            tt[ 2 ] = - (Flt )j / ((Flt )( M - 1 )) ;
            VecCopy( tt, p[ i ][ j ] ) ;
        }
    VecCopy( p11, p[ 1 ][ 1 ] ) ;
    VecCopy( p40, p[ 4 ][ 0 ] ) ;
    VecCopy( p04, p[ 0 ][ 4 ] ) ;
    VecCopy( p33, p[ 3 ][ 3 ] ) ;

    /* calculate spline */
    beta_surface( beta1, beta2, (Point **)p,
                  M, M, DIV, DIV, (Point **)oo ) ;

    /* initialize graphics mode */
    SetVideoMode() ;

    /* initialize 3D graphics */
    Init3D( EyePos, LookAt, Up, Alpha ) ;

    /* display built spline */
    PlotSurface( (Point **)oo, N, N, 1 ) ;
}
```

The source of the program used may be found on the applied diskette in `beta.c` file of the subdirectory `SURFACES`. The detailed information on the file names and their contents is presented in `readme.txt` file of directory `SPLINES` and in Appendix B.

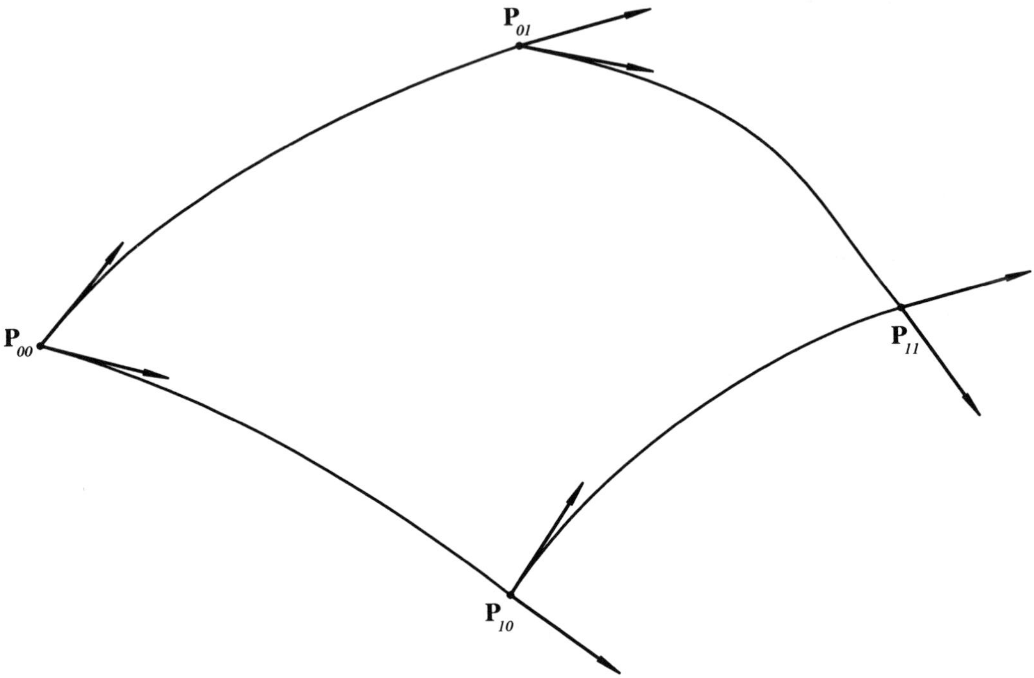

Figure 28.

4.5 Other spline surfaces

4.5.1 Interpolating bicubic Hermite surfaces

For given vertices

$$\mathbf{P}_{00}, \quad \mathbf{P}_{01}, \quad \mathbf{P}_{10}, \quad \mathbf{P}_{11}$$

and a set of vectors

$$\mathbf{X}_{00}, \mathbf{X}_{01}, \mathbf{X}_{10}, \mathbf{X}_{11}, \quad \mathbf{Y}_{00}, \mathbf{Y}_{01}, \mathbf{Y}_{10}, \mathbf{Y}_{11}, \quad \mathbf{Z}_{00}, \mathbf{Z}_{01}, \mathbf{Z}_{10}, \mathbf{Z}_{11}$$

an (elementary) bicubic Hermite surface is defined with the help of the **matrix equation**

$$\mathbf{R}(u,v) = \mathbf{U}^T\mathbf{M}^T\mathbf{G}\mathbf{M}\mathbf{V}, \quad 0 \le u, v \le 1,$$

where

$$\mathbf{R}(u,v) = \begin{pmatrix} x(u,v) \\ y(u,v) \\ z(u,v) \end{pmatrix}, \qquad \mathbf{U} = \begin{pmatrix} u^0 \\ u^1 \\ u^2 \\ u^3 \end{pmatrix}, \qquad \mathbf{V} = \begin{pmatrix} v^0 \\ v^1 \\ v^2 \\ v^3 \end{pmatrix},$$

$$\mathbf{G} = \begin{pmatrix} \mathbf{P}_{00} & \mathbf{P}_{01} & \mathbf{Y}_{00} & \mathbf{Y}_{01} \\ \mathbf{P}_{10} & \mathbf{P}_{11} & \mathbf{Y}_{10} & \mathbf{Y}_{11} \\ \mathbf{X}_{00} & \mathbf{X}_{01} & \mathbf{Z}_{00} & \mathbf{Z}_{01} \\ \mathbf{X}_{10} & \mathbf{X}_{11} & \mathbf{Z}_{10} & \mathbf{Z}_{11} \end{pmatrix}, \quad \mathbf{M} = \begin{pmatrix} 1 & 0 & -3 & 2 \\ 0 & 0 & 3 & -2 \\ 0 & 1 & -2 & 1 \\ 0 & 0 & -1 & 1 \end{pmatrix}.$$

Matrix $\mathbf{M}$ is called *a basic matrix of bicubic Hermite surface* and matrix $\mathbf{G}$ its *geometric matrix.*

Geometric meaning of the vectors defining the elementary bicubic Hermite surface.

Vertices $\mathbf{P}_{ij}$ $(i = 0, 1,\ j = 0, 1)$ are the corner points of the elementary bicubic Hermite surface (Figure 28). Vector $\mathbf{X}_{ij}$ is a tangent vector of the u-line at the point $\mathbf{P}_{ij}$ of the Hermite surface, and vector $\mathbf{Y}_{ij}$ is a tangent vector of the v-line at the point (for regularity of the constructed surface it is natural to require that every pair of vectors $\mathbf{X}_{ij}$ and $\mathbf{Y}_{ij}$ be noncollinear). As for vector $\mathbf{Z}_{ij}$, the vector is a twist vector at vertex $\mathbf{P}_{ij}$.

In particular, at vertex $\mathbf{P}_{00}$ of the elementary bicubic Hermite surface (for $u = 0$ and $v = 0$) we have

$$\mathbf{R}(0,0) = \mathbf{P}_{00},\ \mathbf{R}_u(0,0) = \mathbf{X}_{00},\ \mathbf{R}_v(0,0) = \mathbf{Y}_{00},\ \mathbf{R}_{uv}(0,0) = \mathbf{Z}_{00}.$$

Under an affine transformation of the parameters

$$u \mapsto \xi = (1-u)a + ub, \quad v \mapsto \eta = (1-u)c + ud,$$

when a unit square $[0,1] \times [0,1]$ goes into rectangle $[a,b] \times [c,d]$, the shape of the elementary bicubic Hermite surface changes. To hold the shape it is necessary to replace vectors $\mathbf{X}_{ij}$, $\mathbf{Y}_{ij}$, and $\mathbf{Z}_{ij}$ $(i = 0, \dots, m,\ j = 0, \dots, n)$ with vectors

$$\frac{1}{b-a}\mathbf{X}_{ij}, \quad \frac{1}{d-c}\mathbf{Y}_{ij}, \quad \text{and} \quad \frac{1}{(b-a)(d-c)}\mathbf{Z}_{ij},$$

respectively.

Composite bicubic Hermite surfaces. Curve S is called *a (composite) bicubic Hermite surface* defined by array

$$\begin{array}{lllll} \mathbf{P}_{00}, & \mathbf{P}_{01}, & \dots & \mathbf{P}_{0,n-1}, & \mathbf{P}_{0n}, \\ \mathbf{P}_{10}, & \mathbf{P}_{11}, & \dots & \mathbf{P}_{1,n-1}, & \mathbf{P}_{1n}, \\ \dots & \dots & \dots & \dots & \dots \\ \mathbf{P}_{m-1,0}, & \mathbf{P}_{m-1,1}, & \dots & \mathbf{P}_{m-1,n-1}, & \mathbf{P}_{m-1,n}, \\ \mathbf{P}_{m0}, & \mathbf{P}_{m1}, & \dots & \mathbf{P}_{m,n-1}, & \mathbf{P}_{mn} \end{array}$$

and vectors

$$\mathbf{X}_{kj}, \quad k = 0, m, \quad j = 0, \dots, n; \quad \mathbf{Y}_{il}, \quad i = 0, \dots, m, \quad l = 0, n;$$

$$\mathbf{Z}_{00}, \quad \mathbf{Z}_{0n}, \quad \mathbf{Z}_{m0}, \quad \mathbf{Z}_{mn},$$

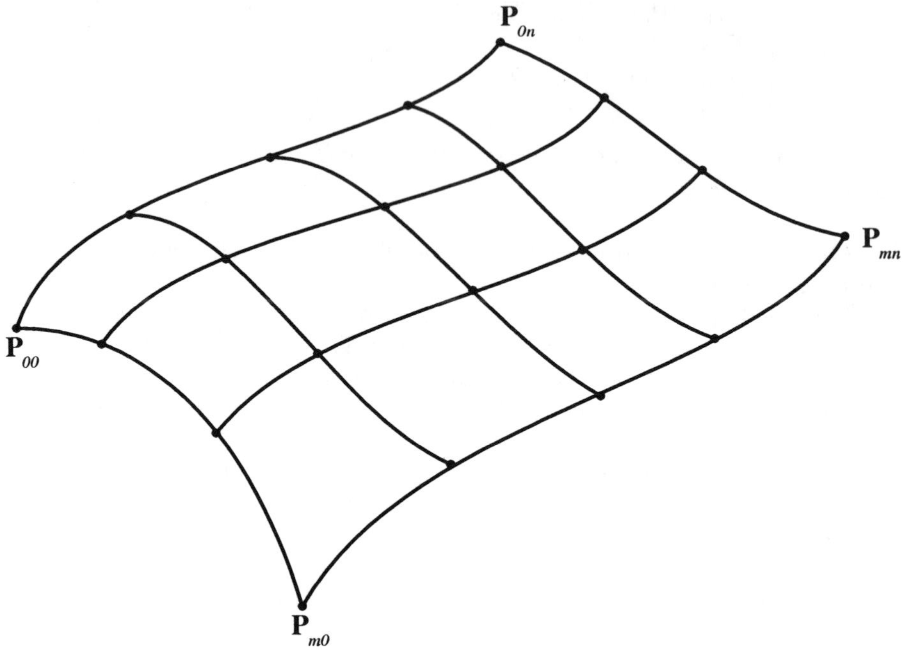

Figure 29.

if the surface may be represented as a union of the elementary bicubic Hermite surfaces $S^{(1,1)}, \ldots, S^{(m,n)}$,

$$S = \cup_{i=1}^{m} \cup_{j=1}^{n} S^{(i,j)}$$

(Figure 29); (i, j)th surface $S^{(i,j)}$ is described by the parametric equation of the form

$$\mathbf{R}^{(i,j)}(u, v) = \mathbf{U}^T \mathbf{M}^T \mathbf{G}^{(i,j)} \mathbf{M} \mathbf{V},$$

$$0 \le u, v \le 1, \quad i = 1, \ldots, m, \quad j = 1, \ldots, n,$$

where $\mathbf{M}$ is the basic matrix of the bicubic Hermite surface and the geometric matrix $\mathbf{G}^{(i,j)}$ has the following form

$$\mathbf{G}^{(i,j)} = \begin{pmatrix} \mathbf{P}_{i-1,j-1} & \mathbf{P}_{i-1,j} & \mathbf{Y}_{i-1,j-1} & \mathbf{Y}_{i-1,j} \\ \mathbf{P}_{i,j-1} & \mathbf{P}_{ij} & \mathbf{Y}_{i,j-1} & \mathbf{Y}_{ij} \\ \mathbf{X}_{i-1,j-1} & \mathbf{X}_{i-1,j} & \mathbf{Z}_{i-1,j-1} & \mathbf{Z}_{i-1,j} \\ \mathbf{X}_{i,j-1} & \mathbf{X}_{ij} & \mathbf{Z}_{i,j-1} & \mathbf{Z}_{ij} \end{pmatrix}.$$

Vectors

$$\mathbf{X}_{ij}, \quad i = 1, \ldots, m-1, \quad j = 0, \ldots, n,$$

$$\mathbf{Y}_{ij}, \quad i = 0, \ldots, m, \quad j = 1, \ldots, n-1,$$

$$\mathbf{Z}_{ij} \quad i \neq 0, m, \quad j \neq 0, n$$

may be chosen in such a way that the composite bicubic Hermite surface is C^2-smooth. All the initial information required for their definition may be written in the form of the following table:

$\mathbf{Z}_{00}$	$\mathbf{X}_{00}$	$\mathbf{X}_{01}$	$\ldots$	$\mathbf{X}_{0,n-1}$	$\mathbf{X}_{0n}$	$\mathbf{Z}_{0n}$
$\mathbf{Y}_{00}$	$\mathbf{P}_{00}$	$\mathbf{P}_{01}$	$\ldots$	$\mathbf{P}_{0,n-1}$	$\mathbf{P}_{0n}$	$\mathbf{Y}_{0n}$
$\mathbf{Y}_{10}$	$\mathbf{P}_{10}$	$\mathbf{P}_{11}$	$\ldots$	$\mathbf{P}_{1,n-1}$	$\mathbf{P}_{1n}$	$\mathbf{Y}_{1n}$
$\ldots$	$\ldots$	$\ldots$	$\ldots$	$\ldots$	$\ldots$	$\ldots$
$\mathbf{Y}_{m-1,0}$	$\mathbf{P}_{m-1,0}$	$\mathbf{P}_{m-1,1}$	$\ldots$	$\mathbf{P}_{m-1,n-1}$	$\mathbf{P}_{m-1,n}$	$\mathbf{Y}_{m-1,n}$
$\mathbf{Y}_{m0}$	$\mathbf{P}_{m0}$	$\mathbf{P}_{m1}$	$\ldots$	$\mathbf{P}_{m,n-1}$	$\mathbf{P}_{mn}$	$\mathbf{Y}_{mn}$
$\mathbf{Z}_{m0}$	$\mathbf{X}_{m0}$	$\mathbf{X}_{m1}$	$\ldots$	$\mathbf{X}_{m,n-1}$	$\mathbf{X}_{mn}$	$\mathbf{Z}_{mn}$

Let us describe the formulas with which these vectors are calculated.

Let us begin with the matrix formulas for the calculation of vectors $\mathbf{X}_{ij}$ and $\mathbf{Y}_{ij}$ at the inner vertices of array $\mathbf{P}$:

$$\begin{pmatrix} 1 & 4 & 1 & & & & \\ & 1 & 4 & 1 & & & \\ & & & \ddots & & & \\ & & & & 1 & 4 & 1 \\ & & & & & 1 & 4 & 1 \end{pmatrix} \begin{pmatrix} \mathbf{X}_{00} & \ldots & \mathbf{X}_{0n} \\ \mathbf{X}_{10} & \ldots & \mathbf{X}_{1n} \\ \vdots & & \vdots \\ \mathbf{X}_{m-1,0} & \ldots & \mathbf{X}_{m-1,0} \\ \mathbf{X}_{m0} & \ldots & \mathbf{X}_{m,0} \end{pmatrix} =$$

$$\begin{pmatrix} -3 & 0 & 3 & & & & \\ & -3 & 0 & 3 & & & \\ & & & \ddots & & & \\ & & & & -3 & 0 & 3 \\ & & & & & -3 & 0 & 3 \end{pmatrix} \begin{pmatrix} \mathbf{P}_{00} & \ldots & \mathbf{P}_{0n} \\ \mathbf{P}_{10} & \ldots & \mathbf{P}_{1n} \\ \vdots & & \vdots \\ \mathbf{P}_{m-1,0} & \ldots & \mathbf{P}_{m-1,0} \\ \mathbf{P}_{m0} & \ldots & \mathbf{P}_{m,0} \end{pmatrix}.$$

(The rectangular scalar matrices have the same size $(m-1) \times (m+1)$, the elements of the first and the last rows of matrix $\mathbf{X}$ are known, and the matrix equation has a unique solution.)

$$\begin{pmatrix} 1 & 4 & 1 & & & & \\ & 1 & 4 & 1 & & & \\ & & & \ddots & & & \\ & & & & 1 & 4 & 1 \\ & & & & & 1 & 4 & 1 \end{pmatrix} \begin{pmatrix} \mathbf{Y}_{00} & \ldots & \mathbf{Y}_{m0} \\ \mathbf{Y}_{01} & \ldots & \mathbf{Y}_{m1} \\ \vdots & & \vdots \\ \mathbf{Y}_{0,n-1} & \ldots & \mathbf{Y}_{0,n-1} \\ \mathbf{Y}_{0n} & \ldots & \mathbf{Y}_{mn} \end{pmatrix} =$$

$$\begin{pmatrix} -3 & 0 & 3 & & & & \\ & -3 & 0 & 3 & & & \\ & & & \ddots & & & \\ & & & & -3 & 0 & 3 \\ & & & & & -3 & 0 & 3 \end{pmatrix} \begin{pmatrix} \mathbf{P}_{00} & \dots & \mathbf{P}_{m0} \\ \mathbf{P}_{01} & \dots & \mathbf{P}_{m1} \\ \vdots & & \vdots \\ \mathbf{P}_{0,n-1} & \dots & \mathbf{P}_{0,n-1} \\ \mathbf{P}_{0n} & \dots & \mathbf{P}_{mn} \end{pmatrix}.$$

(The rectangular scalar matrices have the same size $(n-1) \times (n+1)$, the elements of the first and the last rows of matrix $\mathbf{Y}$ are known, and the matrix equation has a unique solution.)

If vectors $\mathbf{X}_{ij}$ and $\mathbf{Y}_{ij}$ are found, the twist vectors $\mathbf{Z}_{ij}$ at all the inner and corner vertices of array $\mathbf{P}$ (with the exception of four given corner vertices $\mathbf{Z}_{00}$, $\mathbf{Z}_{0n}$, $\mathbf{Z}_{1m}$, $\mathbf{Z}_{mn}$) may be calculated as follows.

First with the help of formulas

$$\begin{pmatrix} 1 & 4 & 1 & & & & \\ & 1 & 4 & 1 & & & \\ & & & \ddots & & & \\ & & & & 1 & 4 & 1 \\ & & & & & 1 & 4 & 1 \end{pmatrix} \begin{pmatrix} \mathbf{Z}_{k0} \\ \mathbf{Z}_{k1} \\ \vdots \\ \mathbf{Z}_{k,n-1} \\ \mathbf{Z}_{kn} \end{pmatrix}$$

$$= \begin{pmatrix} -3 & 0 & 3 & & & & \\ & -3 & 0 & 3 & & & \\ & & & \ddots & & & \\ & & & & -3 & 0 & 3 \\ & & & & & -3 & 0 & 3 \end{pmatrix} \begin{pmatrix} \mathbf{X}_{k0} \\ \mathbf{X}_{k1} \\ \vdots \\ \mathbf{X}_{k,n-1} \\ \mathbf{X}_{kn} \end{pmatrix}, \qquad k = 0, m$$

we calculate the twist vectors at the boundary vertices $\mathbf{P}_{kj}$ $(j = 1, \dots, n-1)$ (the rectangular scalar matrices have the same size $(n-1) \times (n+1)$, the first and the last elements of column $\mathbf{Z}$ are known, and the matrix equation has a unique solution) and then with the help of formulas

$$\begin{pmatrix} 1 & 4 & 1 & & & & \\ & 1 & 4 & 1 & & & \\ & & & \ddots & & & \\ & & & & 1 & 4 & 1 \\ & & & & & 1 & 4 & 1 \end{pmatrix} \begin{pmatrix} \mathbf{Z}_{0j} \\ \mathbf{Z}_{1j} \\ \vdots \\ \mathbf{Z}_{m-1,j} \\ \mathbf{Z}_{mj} \end{pmatrix}$$

$$= \begin{pmatrix} -3 & 0 & 3 & & & & \\ & -3 & 0 & 3 & & & \\ & & & \ddots & & & \\ & & & & -3 & 0 & 3 \\ & & & & & -3 & 0 & 3 \end{pmatrix} \begin{pmatrix} \mathbf{Y}_{0j} \\ \mathbf{Y}_{1j} \\ \vdots \\ \mathbf{Y}_{m-1,j} \\ \mathbf{Y}_{mj} \end{pmatrix}, \qquad l = 0, n$$

we calculate the twist vectors at the boundary vertices $\mathbf{P}_{il}$ $(i = 1, \ldots, m-1)$ (the rectangular scalar matrices have the same size $(m-1) \times (m+1)$, the first and the last elements of column $\mathbf{Z}$ are known, and the matrix equation has a unique solution).

By doing so, we completely construct the arrays

of the radius-vector derivatives along the u-lines

$$\begin{array}{lllll} \mathbf{X}_{00} & \mathbf{X}_{01} & \ldots & \mathbf{X}_{0,n-1} & \mathbf{X}_{0n} \\ \mathbf{X}_{10} & \mathbf{X}_{11} & \ldots & \mathbf{X}_{1,n-1} & \mathbf{X}_{1n} \\ \ldots & \ldots & \ldots & \ldots & \ldots \\ \mathbf{X}_{m-1,0} & \mathbf{X}_{m-1,1} & \ldots & \mathbf{X}_{m-1,n-1} & \mathbf{X}_{m-1,n} \\ \mathbf{X}_{m0} & \mathbf{X}_{m1} & \ldots & \mathbf{X}_{m,n-1} & \mathbf{X}_{mn} \end{array}$$

of the radius-vector derivatives along the v-lines

$$\begin{array}{lllll} \mathbf{Y}_{00} & \mathbf{Y}_{01} & \ldots & \mathbf{Y}_{0,n-1} & \mathbf{Y}_{0n} \\ \mathbf{Y}_{10} & \mathbf{Y}_{11} & \ldots & \mathbf{Y}_{1,n-1} & \mathbf{Y}_{1n} \\ \ldots & \ldots & \ldots & \ldots & \ldots \\ \mathbf{Y}_{m-1,0} & \mathbf{Y}_{m-1,1} & \ldots & \mathbf{Y}_{m-1,n-1} & \mathbf{Y}_{m-1,n} \\ \mathbf{Y}_{m0} & \mathbf{Y}_{m1} & \ldots & \mathbf{Y}_{m,n-1} & \mathbf{Y}_{mn} \end{array}$$

of the twist vectors

$$\begin{array}{lllll} \mathbf{Z}_{00} & \mathbf{Z}_{01} & \ldots & \mathbf{Z}_{0,n-1} & \mathbf{Z}_{0n} \\ \mathbf{Z}_{10} & \mathbf{Z}_{11} & \ldots & \mathbf{Z}_{1,n-1} & \mathbf{Z}_{1n} \\ \ldots & \ldots & \ldots & \ldots & \ldots \\ \mathbf{Z}_{m-1,0} & \mathbf{Z}_{m-1,1} & \ldots & \mathbf{Z}_{m-1,n-1} & \mathbf{Z}_{m-1,n} \\ \mathbf{Z}_{m0} & \mathbf{Z}_{m1} & \ldots & \mathbf{Z}_{m,n-1} & \mathbf{Z}_{mn}. \end{array}$$

The closing step in construction of the composite bicubic Hermite surface is the calculation of radius-vectors $\mathbf{R}^{(i,j)}(u, v)$ of the elementary fragments $S^{(i,j)}$ by formula

$$\mathbf{R}^{(i,j)}(u, v) = \mathbf{U}^T \mathbf{M}^T \mathbf{G}^{(i,j)} \mathbf{M} \mathbf{V},$$

$$0 \le u, v \le 1, \quad i = 1, \ldots, m, \quad j = 1, \ldots, n.$$

Properties of the composite bicubic Hermite surface. A composite bicubic Hermite surface generated by array

$$\mathbf{P} = \{\mathbf{P}_{ij}, \quad i = 0, \ldots, m, j = 0, \ldots, n\}$$

and vectors

$$\mathbf{X}_{kj}, \quad k = 0, m, \quad j = 0, \ldots, n,$$

$$\mathbf{Y}_{il}, \quad i = 0, \ldots, m, \quad l = 0, n,$$

$$\mathbf{Z}_{00}, \quad \mathbf{Z}_{0n}, \quad \mathbf{Z}_{m0}, \quad \mathbf{Z}_{mn}$$

possesses the following properties:

1^+ the surface is a C^2-smooth surface (the surface with continuous curvature).

2^+ the surface interpolates the vertices of array $\mathbf{P}$.

3^+ the tangent vectors $\mathbf{X}_{ij}$ and $\mathbf{Y}_{ij}$ to the coordinate lines u = const and v = const of the composite surface S at the inner vertex $\mathbf{P}_{ij}$ ($i = 1, \ldots, m-1$, $j = 1, \ldots, n-1$) as well as the twist vectors $\mathbf{Z}_{ij}$ are uniquely determined by the vertices of array $\mathbf{P}$ and vectors $\mathbf{X}_{ij}$, $\mathbf{Y}_{ij}$, and $\mathbf{Z}_{ij}$ given at the boundary vertices.

4^- the surface does not lie in the convex hull generated by the given array.

5^- the change of one vertex in the given array or one of the given vectors causes the whole surface shape to change.

6^- with addition of only one vertex to the given array, it is necessary to recalculate all the parametric equations.

7^+ the composite bicubic Hermite surface is affine invariant.

8^- the equations of the composite bicubic Hermite surface do not contain free parameters (because of this, the surface is uniquely determined by the given array of vertices and the tangent vector set, and there is no way of controlling its shape).

9^- the composite bicubic Hermite surface is projective-noninvariant.

4.5.2 Implementation

The following C function is an implementation of the algorithm described above.

Function syntax:

```
hermite_surface( p, x, y, z, m, n, div_x, div_y, output )
```

Function parameters are:

```
Point **p        (in)   - spline control net
                          points Pij, i = 0...(m-1),
                                      j = 0...(n-1)
Point **x        (in)   - array of u-derivatives
                          elements Xij, i = 0...(m-1),
                                        j = 0...(n-1)
                          must be filled:
                          Xkj, k = 0, m-1, j = 0 ... n-1
Point **y        (in)   - array of v-derivatives
                          elements Yij, i = 0...(m-1),
                                        j = 0...(n-1)
                          must be filled:
                          Yil, i = 0...m-1, l = 0, n-1
```

```
 Point **z       (in)  - array of twist vectors
                         elements Zij, i = 0...(m-1),
                                       j = 0...(n-1)
                         must be filled:
                         Z[0][0], Z[0][n-1],
                         Z[m-1][0], Z[m-1][n-1]
 int m, n        (in)  - number of points in control net
 int div_x       (in)  - number of links in "u-polyline"
 int div_y       (in)  - number of links in "v-polyline"
 Point **output (out) - spline approximating net
                         points Qij, i = 0...(m-1)*div_x
                                     j = 0...(n-1)*div_y
```

The example below shows how to use this function to build a spline surface:

```
#include "lib.h"

/* Imported function declaration */
void hermite_surface( Point **p, Point **x, Point **y,
                      Point **z, int m, int n,
                      int div_x, int div_y,
                      Point **output ) ;

/* number of points in control graph */
#define  M    (5)
#define DIV   (5)
/* number of points in approximating net */
#define  N   ((M-1)*DIV+1)

/* Control graph initialization */
Point p[ M ][ M ] ;
Point p11 = { .25, 1, -.25  } ;
Point p40 = { 1.0, -.5, 0.0 } ;
Point p04 = { 0.0, 2, -1.0  } ;
Point p33 = { .75, 2, -.75  } ;

Point x[ M ][ M ] ;
Point y[ M ][ M ] ;
Point z[ M ][ M ] ;

/* auxiliary variable */
Point tt = { 0, 0, 0 } ;
Point xx = { 1, 0, 0 } ;
Point yy = { 0, 0, -1 } ;
Point zz = { 0, 0, 0 } ;
```

```
/* Declaration of output arrays */
Point oo[ N ][ N ] ;

Point EyePos = { .5, 3, 3 } ;
Point LookAt = { .5, 0, -5 } ;
Vec Up = { 0, 1, -1 } ;
Flt Alpha = .7 ;

void main( void )
{
    int i, j ;

    /* Point array fill */
    for( i = 0; i < M; i ++ )
        for( j = 0; j < M; j ++ )
        {
            tt[ 0 ] = (Flt )i / ((Flt )( M - 1 )) ;
            tt[ 2 ] = - (Flt )j / ((Flt )( M - 1 )) ;
            VecCopy( tt, p[ i ][ j ] ) ;
        }
    VecCopy( p11, p[ 1 ][ 1 ] ) ;
    VecCopy( p40, p[ 4 ][ 0 ] ) ;
    VecCopy( p04, p[ 0 ][ 4 ] ) ;
    VecCopy( p33, p[ 3 ][ 3 ] ) ;

    for( j = 0; j < M; j ++ )
    {
        VecCopy( xx, x[  0  ][ j ] ) ;
        VecCopy( xx, x[ M-1 ][ j ] ) ;
    }
    for( i = 0; i < M; i ++ )
    {
        VecCopy( y[ i ][  0  ], yy ) ;
        VecCopy( y[ i ][ M-1 ], yy ) ;
    }
    VecCopy( zz, z[  0  ][  0  ] ) ;
    VecCopy( zz, z[  0  ][ M-1 ] ) ;
    VecCopy( zz, z[ M-1 ][  0  ] ) ;
    VecCopy( zz, z[ M-1 ][ M-1 ] ) ;

    /* calculate spline */
    hermite_surface( (Point **)p, (Point **)x,
                     (Point **)y, (Point **)z,
                     M, M, DIV, DIV, (Point *)oo ) ;
```

```
    /* initialize graphics mode */
    SetVideoMode() ;

    /* initialize 3D graphics */
    Init3D( EyePos, LookAt, Up, Alpha ) ;

    /* display built spline */
    PlotSurface( (Point **)oo, N, N, 1 ) ;
}
```

The source of the program used may be found on the applied diskette in `hermite.c` file of the subdirectory `SURFACES`. The detailed information on the file names and their contents is presented in `readme.txt` file of directory `SPLINES` and in Appendix B.

4.5.3 Implicit cubic spline surfaces

In last few years, the works devoted to composite surfaces, elementary fragments of which are described by implicit equations of the form

$$F(x, y, z) = 0,$$

where $F(x, y, z)$ is an algebraic polynomial of the second or third degree, have appeared. Especially great interest is displayed by the users of computer graphics. This mode of description is very convenient for generating realistic three-dimensional images by the ray-tracing method.

It is an interesting but not simple question. In this handbook there is no opportunity to discuss this question more closely so we refer readers to special publications (for details see some of them at the end of the book).

Appendix A

Programs of the sweep method for tridiagonal and pentadiagonal matrices

Every so often when constructing certain spline functions, spline curves, and spline surfaces there is a need to find solutions of the linear algebraic systems with the sparse matrices. In this appendix we describe programs (in Fortran) for solving the linear systems with tridiagonal (`Progon3` subsection A) and pentadiagonal (`Progon5` subsection B) matrices by the sweep method.

A. Tridiagonal matrix

```
        subroutine Progon3(a,b,c,d,u,v,w,s,t,x,n)
c**********************************************************
c       Program Progon3 for solving the linear system    *
c       of the form                                       *
c       a(1)*x(1) + b(1)*x(2)      + c(1)*x(n) = d(1)     *
c       .........................................        *
c       c(i)*x(i-1) + a(i)*x(i)  + b(i)*x(i+1) = d(i)     *
c       .........................................        *
c       b(n)*x(1)     + a(n)*x(n-1) + b(n)*x(n)= d(n)     *
c       with tridiagonal matrix by sweep method.          *
c   n - number of equations                               *
c a,b,                                                    *
c c,d  - arrays (of size n) containing elements of the    *
c        matrix diagonals and of the right parts          *
c  w,s,t,                                                 *
c  u,v - work arrays (of size n+1)                        *
c Output:                                                 *
c x - array (of size n) containing solution of the        *
c     system                                              *
c**********************************************************
        dimension a(1),b(1),c(1),d(1),u(1),
     *             v(1),x(1),w(1),s(1),t(1)
c
        u(1) = 0.
        v(1) = 0.
        w(1) = 1.
c
        do  i = 1,n
            i1 = i + 1
            z  = 1./(a(i) + c(i)*v(i))
            v(i1) = -b(i)*z
            u(i1) = (-c(i)*u(i) + d(i))*z
            w(i1) = - c(i)*w(i)*z
        enddo
c
    s(n) = 1.
    t(n) = 0.
    do  i = n-1,1, -1
        s(i) = v(i+1)*s(i+1) + w(i+1)
        t(i) = v(i+1)*t(i+1) + u(i+1)
    enddo
c
    x(n) = (d(n) - b(n)*t(1) - c(n)*t(n-1))/
```

```
     *            (a(n) + b(n)*s(1) + c(n)*s(n-1))
     do i = 1, n-1
         x(i) = s(i)*x(n) + t(i)
     enddo
c
          return
          end
```

B. Pentadiagonal matrix

```
        subroutine Progon5(n,a,b,c,d,e,g,u,v,w,
     *                       p,q,r,s,t,aa,dd,z)
c
c*********************************************************
c Program Progon5 for solving linear system of the       *
c form                                                   *
c a1*x1+b1*x2+c1*x3+            e1*x(n-1)+d1*x(n) = g1   *
c d2*x1+a2*x2+b2*x3+c2*x4+                e2*x(n) = g2   *
c e3*x1+d3*x2+a3*x3+b3*x4+c4*x4                   = g3   *
c       e4*x2+d4*x3+a4*x4+b4*x5+c5*x5             = g4   *
c .............................................          *
c bn*x1+cn*x2+.......+en*x(n-2)+dn*x(n-1)+an*x(n) = g(n) *
c with pentadiagonal matrix by sweep method              *
c n - number of equations                                *
c a,b,c,d,e - arrays (of size n) for elements of         *
c             the matrix diagonals                       *
c g - array (of size n) for elements of the right parts  *
c u,v,w,p,q,                                             *
c r,s,t,aa,dd - work arrays (of size n+2)                *
c    Output:                                             *
c z - array (of size n) containing solution of the       *
c    system                                              *
c*********************************************************
c
       real a(1),b(1),c(1),d(1),e(1),g(1),u(1),aa(1),r(1)
       real v(1),w(1),p(1),q(1),t(1),s(1),z(1),dd(1)
c
             v(1) = 0.
             v(2) = 0.
             w(1) = 0.
             w(2) = 0.
             u(1) = 0.
             u(2) = 0.
c
        do i = 1,n
```

```
         dd(i)   =  d(i) + e(i)*v(i)
         aa(i)   =  a(i) + dd(i)*v(i+1) +e(i)*w(i)
         u(i+2) = (g(i) - dd(i)*u(i+1) - e(i)*u(i))/aa(i)
         v(i+2) = -(b(i)+dd(i)*w(i+1))/aa(i)
         w(i+2) = -c(i)/aa(i)
      enddo
c
            p(1) = 0.
            q(2) = 0.
            p(2) = 1.
            q(1) = 1.
c
      do i = 1,n
         p(i+2) = -(dd(i)*p(i+1) + e(i)*p(i))/aa(i)
         q(i+2) = -(dd(i)*q(i+1) + e(i)*q(i))/aa(i)
      enddo
c
            t(n-1) = 0.
            s(n-1) = 0.
            t(n)   = 0.
            r(n)   = 0.
            r(n-1) = 1.
            s(n)   = 1.
c
      do i = n-2,1,-1
         t(i) = v(i+2)*t(i+1) +w(i+2)*t(i+2) + u(i+2)
         s(i) = v(i+2)*s(i+1) +w(i+2)*s(i+2) + p(i+2)
         r(i) = v(i+2)*r(i+1) +w(i+2)*r(i+2) + q(i+2)
      enddo
c
      a11 =  1. - q(n+1) - w(n+1)*r(1)
      a12 = -(p(n+1) + v(n+1) + w(n+1)*s(1))
      a21 = -(v(n+2)*r(1) + w(n+2)*r(2) + q(n+2))
      a22 =  1. - p(n+2) - v(n+2)*s(1) - w(n+2)*s(2)
      b1  = w(n+1)*t(1) + u(n+1)
      b2  = v(n+2)*t(1) + w(n+2)*t(2) + u(n+2)
      z(n-1) = ( b1*a22 - b2*a12)/(a11*a22-a12*a21)
      z(n)   = (-b1*a21 + b2*a11)/(a11*a22-a12*a21)
c
      do i = 1,n-2
         z(i) = t(i) + s(i)*z(n) + r(i)*z(n-1)
      enddo
c
      return
      end
```

Appendix B

Description of diskette

SplineGuide

Programs in Fortran and C for:

interpolating cubic spline functions of one variable,
smoothing cubic spline functions of one variable,
interpolating bicubic spline functions of two variables,
smoothing bicubic spline functions of two variables,
constructing cubic Bézier curves,
constructing cubic B-spline curves,
constructing Beta-spline curves,
constructing cubic Hermite curves,
constructing Catmull-Rom curves,
constructing bicubic Bézier surfaces,
constructing bicubic B-spline surfaces,
constructing Beta-spline surfaces,
constructing bicubic Hermite surfaces.

For each program, one or more examples of its operation are offered.

SplineGuide is designed to be extremely user-friendly and built-in help is easily accessible.

SplineGuide is designed to be used with an IBM PC XT/AT or compatible computer with an EGA/VGA color or monochrome monitor, may be stored on one floppy disc, and does not require any additional memory or co-processor.

List of files on disc

```
\SPLINES\readme.txt
        APPENDIX\progon3.for
                 progon5.for

       \FUNCTION\ONEVARIA\INTERPOL\spline.for
                                   exmpl1.for
                                   exmpl2.for
                                   exmpl3.for
                                   exmpl4.for
                                   res1.dat
                                   res2.dat
                                   res3.dat
                                   res4.dat
                         \SMOOTH\smspline.for
                                 exmpl1.for
                                 exmpl2.for
                                 exmpl3.for
                                 res1.dat
                                 res2.dat
                                 res3.dat
                                 input1.dat
                                 input2.dat
                                 input3.dat
                \TWOVARIA\INTERPOL\splint2.for
                                   exmpl1.for
                                   exmpl2.for
                                   exmpl3.for
                                   exmpl4.for
                                   res1.dat
                                   res2.dat
                                   res3.dat
                                   res4.dat
                         \SMOOTH\smspl2.for
                                 exmpl1.for
                                 exmpl2.for
                                 exmpl3.for
                                 res1.dat
                                 res2.dat
                                 res3.dat


\SPLINES\GEOMETRY\CURVES\BETA\beta.c
                              test.c
```

```
              test.doc
        \BEZIER\bezier.c
                test.c
                test.doc
        \BSPLINE\bspline.c
                 test.c
                 test.doc
        \CATMULL\catmull.c
                 test.c
                 test.doc
        \HERMITE\hermite.c
                 test.c
                 test.doc
        \NURBS\nurbs.c
               test.c
               test.doc
        \global.doc

\FINDC1\findc1.c

\LIB\comon.h
    \lib.c
    \lib.h

\SURFACES\BETA\beta.c
               test.c
               test.doc
         \BEZIER\bezier.c
                 test.c
                 test.doc
         \HERMITE\hermite.c
                  test.c
                  test.doc
         \NURBS\nurbs.c
                test.c
                test.doc
         \global.doc
```

File descriptions

1. **Files** `lib.h, lib.c, common.h, test.c, test.doc.`

File `lib.h` contains declarations of auxiliary functions:

`DrawPolygon()` – displays a polygon on the screen,
`Init3D()` – initializes parameters and vectors required for 3D graphics,
`PlotSurface()` – displays a surface from 2D array on the screen,
`SetVideoMode()` – initializes graphic mode with resolution 640 by 480 and with 16 colors.

File `lib.c` contains implementation of auxiliary functions declared in `lib.h`. It can be used with Microsoft C or BORLAND C compilers without any changes, however it should be modified to be used with other compilers.

File `common.h` defines all necessary types, vector maintenance macros and 2D array access macros.

File `test.c` runs specific test according to spline file (see 2 below).

File `test.doc` describes the syntax of using spline function and contains a corresponding example.

2. **Spline files. Directory** `FUNCTION`.

All the files `*.for` contain source programs in Fortran, and can be used with Microsoft Fortran 5.1 or Microsoft Fortran Power Station compilers.

File `spline.for` (from subdirectory `ONEVARIA`) contains source program `Spline` to calculate an interpolating cubic spline.

File `exmpl1.for` contains main program illustrating use of program `Spline` to construct an interpolating cubic spline for the end conditions of the first type. See file `res1.dat` for results. This example is described on page 30.

File `exmpl2.for` contains main program illustrating use of program `Spline` to construct an interpolating cubic spline for the end conditions of the second type. See file `res2.dat` for results.

File `exmpl3.for` contains main program illustrating use of program `Spline` to construct an interpolating cubic spline for the end conditions of the third type. See file `res3.dat` for results.

File `exmpl4.for` contains main program illustrating use of program `Spline` to construct an interpolating cubic spline for the end conditions of the fourth type. See file `res4.dat` for results.

File `smspline.for` (from subdirectory `ONEVARIA`) contains source program `Smspline` to calculate a smoothing cubic spline.

File `exmpl1.for` contains main program illustrating use of program `Smspline` to construct a smoothing cubic spline for the end conditions of the first type. Initial data for this example are in file `input1.dat`. See file `res1.dat` for results. This example is described on page 39.

File `exmpl2.for` contains main program illustrating use of program `Smspline` to construct a smoothing cubic spline for the end conditions of the second type. Initial data for this example are in file `input2.dat`. See file `res2.dat` for results.

File `exmpl3.for` contains main program illustrating use of program `Smspline` to construct a smoothing cubic spline for the end conditions of the third type. Initial data for this example are in file `input3.dat`. See file `res3.dat` for results.

File `splint2.for` (from subdirectory `TWOVARIA`) contains source program `Splint2` to calculate an interpolating bicubic spline.

File `exmpl1.for` contains main program illustrating use of program `Splint2` to construct an interpolating bicubic spline for the boundary conditions of the first type. See file `res1.dat` for results. This example is described on page 61.

File `exmpl2.for` contains main program illustrating use of program `Splint2` to construct an interpolating bicubic spline for the boundary conditions of the second type. See file `res2.dat` for results.

File `exmpl3.for` contains main program illustrating use of program `Splint2` to construct an interpolating bicubic spline for the boundary conditions of the third type. See file `res3.dat` for results.

File `exmpl4.for` contains main program illustrating use of program `Splint2` to construct an interpolating bicubic spline for the boundary conditions of the fourth type. See file `res4.dat` for results.

File `smspl2.for` (from subdirectory `TWOVARIA`) contains source program `Smspl2` to calculate a smoothing bicubic spline.

File `exmpl1.for` contains main program illustrating use of program `Smspl2` to construct a smoothing bicubic spline for the boundary conditions of the first type. See file `res1.dat` for results. This example is described on page 69.

File `exmpl2.for` contains main program illustrating use of program `Smspl2` to construct a smoothing bicubic spline for the boundary conditions of the second type. See file `res2.dat` for results.

File `exmpl3.for` contains main program illustrating use of program `Smspl2` to construct a smoothing bicubic spline for the boundary conditions of the third type. See file `res3.dat` for results.

3. **Spline files. Directory** `GEOMETRY`.

File `beta.c` (from subdirectory `CURVES`) contains C function to calculate approximating Beta-spline curve for a given 2D data set.

File `beta.c` (from subdirectory `SURFACES`) contains C function to calculate approximating Beta-spline surface for a given 3D data set.

File `bezier.c` (from subdirectory `CURVES`) contains C function to calculate approximating cubic Bézier curve for a given 2D data set.

File `bezier.c` (from subdirectory `SURFACES`) contains C function to calculate approximating bicubic Bézier surface for a given 3D data set.

File `bspline.c` (from subdirectory `CURVES`) contains C function to calculate approximating cubic B-spline curve for a given 2D data set.

File `bspline.c` (from subdirectory `SURFACES`) contains C function to calculate approximating bicubic B-spline surface for a given 3D data set.

File `catmull.c` (from subdirectory `CURVES`) contains C function to calculate approximating Catmull-Rom spline curve for a given 2D data set.

File `hermite.c` (from subdirectory `CURVES`) contains C function to calculate approximating cubic Hermite curve for a given 2D data set.

File `hermite.c` (from subdirectory `SURFACES`) contains C function to calculate approximating bicubic Hermite surface for a given 3D data set.

File `nurbs.c` (from subdirectory `CURVES`) contains C function to calculate approximating nonuniform rational cubic B-spline curve for a given 2D data set.

File `nurbs.c` (from subdirectory `SURFACES`) contains C function to calculate approximating nonuniform rational bicubic B-spline surface for a given 3D data set.

4. **Other files.**

File `global.doc` (from subdirectories `CURVES` and `SURFACE`) contains information concerning use of spline functions.

File `findc1.doc` (from subdirectory `FINDC1`) contains C function to calculate C^1-smooth composite 3D cubic Bézier curve.

File `progon3.for` (from subdirectory `APPENDIX`) contains program (in `Fortran`) of sweep method for linear algebraic systems with tridiagonal matrix.

File `progon5.for` (from subdirectory `APPENDIX`) contains program (in `Fortran`) of sweep method for linear algebraic systems with pentadiagonal matrix.

File `readme.txt` contains description of the applied diskette.

Bibliography

Classical works

Weierstrass K., Sitzungsber. Acad. Berlin 1885 S.633–639, 789–805.

Runge C., Ueber empirishe Funktionen und die Interpolation zwischen aequidistanten Ordinaten. ZAMM, 46:224–243, 1901.

Bernstein S., Démonstration du théorème de Weierstrass fondeé sur le calcul des probabilités. Harkov Soobs. Matem. ob-va, 13:1–2, 1912.

Theoretical works on splines

Ahlberg J., Nilson E., Walsh J., The theory of splines and their applications. Academic Press, 1967.

Stechkin S.B., Subbotin Yu.N., Splines in computer mathematics. Nauka, 1976 (in Russian).

Zavjalov Yu.S., Kvasov B.I., Miroshnichenko V.L., Methods of spline functions. Nauka, 1980 (in Russian).

Splines in computer graphics

Barsky B., Computer Graphics and Geometric Modeling using Beta-splines. Springer Verlag, 1988.

Bartles R., Beatty J., Barsky B., An introduction to splines for use in computer graphics and geometric modeling. Morgan Kaufmann, 1987.

Farin G., Curves and surfaces for computer aided geometric design. A practical guide. Academic Press, 1990.

Foley J.D., van Dam A., Feiner S.K., Hugues J.F., Computer graphics. Principles and practice. Addison-Wesley Pub.Com,1991.

Splines and design

Faux I.D., Pratt M.J., Computational geometry for design and manufacture. Ellis Horwood, 1979.

Mathématique et CAO (in four volumes). Hermes Publishing, 1986, 1987.

Zavjalov Yu.S., Leus V.A., Skorospelov B.A., Splines in the engineering geometry. Mashinostroenie, 1985 (in Russian).

Yamaguchi F., Curves and surfaces in computer aided geometric design. Springer Verlag, 1988.

Practical guides

de Boor C., A practical guide to splines. Springer, 1978.

Recent works on implicit cubic splines

Sederberg T.W. Techniques for cubic algebraic surfaces. IEEE Computer Graphics and Applications, 4 (1990), 14–25.

Sederberg T.W. Techniques for cubic algebraic surfaces. IEEE Computer Graphics and Applications, 5 (1990), 12–21.

Bajaj C.L., Surface fitting using implicit algebraic surface patches. In Topics in Surface Modeling, H.Hagen, Ed., SIAM, 1992.

Paluszny M., Patterson R.R. A family of tangent continuous cubic algebraic splines. Transactions on Graphics, volume 12, N 3 (1993), 209–232.

Dahmen W., Thamm-Schaar T.-M. Cubicoids: modeling and visualization, Computer Aided Geometric Design, 10 (1993), 89–108.

Guo B., Nonsplitting macro patches for implicit cubic spline surfaces, EUROGRAPHCS'93, volume 12, N 3 (1993), C-433 –C-445.

Index

S

T

U

W